How To Get Water Smart

How To Get Water Smart

Products and Practices For Saving Water In The Nineties

Buzz Buzzelli
Peggy Good
Janice S. McCormick
John R. McCormick

FIRMA™
PUBLISHING
Santa Barbara, California
1991

Readers please note:

This book is intended to provide the authors' opinions in regards to the subject matter covered. It is sold with the understanding that neither the authors nor the publisher are engaged in rendering engineering, landscaping, plumbing, wiring, contracting or other professional services. If these services are needed, a competent professional should be sought.

The authors and the publisher specifically disclaim any personal liability, loss, or risk, personal or otherwise, incurred as a consequence, directly or indirectly, of the use and application of any advice or information contained in this book. All product and statistical information has been presented as accurately as possible, although errors in fact may still occur. The authors and publisher disclaim any liability due to such errors.

ISBN 0-9628895-0-4

Library of Congress Catalog Card Number: 91-65066

First edition

10 9 8 7 6 5 4 3 2 1

Terra Firma Publishing
P.O. Box 91315, Santa Barbara, CA 93190-1315
(805) 962-0962

Cover art by Russell Hodin   Book design by The Lily Guild

Manufactured in the United States of America
Printed on recycled paper

For our blue planet
and all of its inhabitants –
may there always be
enough water for us all.

Contents

Foreword

Welcome to *How to Get Water Smart.* This book is intended to serve as a comprehensive, user-friendly guide to saving water in and around the home. By using the word "home," of course, we mean to include houses, apartments, condominiums, townhouses, cabins, cottages, mobile homes, and maybe even caves. Wherever you live, this book is designed to help you save water.

For ease of reference, *How to Get Water Smart* is divided into five chapters, each dealing with a primary water-use area of the home. Some of these areas – such as the bathroom – are common to most every home, so the products and practices contained therein will serve a broad purpose. On the other hand, we realize that not everybody has a yard with five acres, 37 fruit trees and an Olympic-sized pool. But there's information for these citizens too. The rest of us will probably read it anyway...and wonder how you're going to install drip lines around all 37 trees.

No amount of pages could include all of the various products and practices that are available to save water. In the interests of brevity we've selected the ones that we feel have the most merit, or are the most enjoyable or interesting, or that represent a new approach. In almost every arena, you will find a wealth of local information that can guide and help you along once you've embraced the concept of a water-smart home.

If *How to Get Water Smart* is to succeed, it will do so by opening your eyes to the possibilities for simply managing water use in each area of your home. So use this book as a reference volume or just browse. It has been designed to provide useful information on every page.

Introduction

The United States was one of three countries listed by *Time* magazine in which 100 percent of the population has access to safe drinking water. This reassuring little fact is immensely important to us, but it's probably also a fact that ranks fairly low on our list of daily concerns. Secure in our land of plenty, clean abundant water has always seemed like an inalienable right.

But is it really? It seems that everywhere water is becoming more important as an issue – and as an economic and political tool. Many of us are just beginning to learn that drought isn't the only circumstance under which water shortages can occur. Population growth or migration can also cause shortages by overburdening water supplies that have traditionally proved abundant.

Throughout the western states during the 20th century, lives and livelihoods have been shaped at least in part by a largely invisible water issue. The mountain states, with their fragile ecologies and fickle winter snowfall, have always been subject to the vagaries of the seasons. And at the time of this writing, city officials in some parts of Georgia and Florida are experiencing water-availability concerns as growth continues.

When the most recent cyclical drought reached a crisis level in Southern California in 1990 and 1991, various local governments responded by asking citizens to cut water use by 10 to 45 percent, sometimes with stiff rate penalties for noncompliance. For a region used to abundance and green lawns, water conservation seemed a bitter pill indeed. What happened to nirvana?

Residents quickly found that water conservation around the home was more than a patriotic duty – it was an economic necessity. What started out as a call to arms quickly became the aqueous version of paying taxes...on a daily basis.

To cope with the shortfall, we clumsily carried buckets with us into the shower, then even more clumsily lugged them outdoors to pour the grey water on our browning bluegrass. We left toilets unflushed at the risk of domestic friction. And when we went to the movies, the sight of a faucet running on screen moved the audience to cry out in a chorus, "Turn it off!"

Snapped into focus and spurred into action through necessity, we quickly became obsessed with water conservation, and the obsession wasn't fun. The reason why was obvious to those who stopped to look: The awkwardness of buckets and kitchen timers governing our daily rituals quickly wore us down. We didn't know it at the time, but what we needed was a sustainable action plan for smart water use...a water-smart system that we could operate with a minimum of effort – and without driving us crazy in the process. What we needed was a new level of control over the way we used water, our most precious resource.

That is precisely the mission of *How to Get Water Smart.* Filling these pages are user-friendly products and practices that make saving water effortless – or at least more natural. The wide array of water-saving products and practices is intended to allow you to select both the hardware and the methodologies that fit your home, budget and lifestyle.

Hopefully, you will find the concepts useful as well as enjoyable. For along with saving water, you can rejoice in saving the energy necessary to dam it or pump it out of the ground, purify and transport it. You will reduce pressure on the environ-ments from which water is removed before it is delivered to your home. And finally, you may either immediately or ultimately save money, because a part of your monthly budget invariably goes toward water.

Thank you for selecting this book to help plan a more water-efficient home. But let's not end the relationship here – we would like to hear some of your thoughts or experiences regarding water conservation. If you would like to share any comments with us, please direct them to Terra Firma Publishing, P.O. Box 91315, Santa Barbara, CA 93190-1315.

John L. Stein, Publisher

Summer 1991

Bathroom

Like the aroma of freshly brewed coffee, the first pleasurable splash of a warm morning shower lures us out from under the covers to face the day's challenges. Little wonder that up to 75 percent of the water used in homes is consumed in the bathroom.

By far, toilet flushing accounts for the largest percentage of water used in the bathroom – a whopping 50 percent. Add to this the fact that one in five toilets in America leaks up to four gallons each day while not in operation. All told, water consumed by toilets adds up to 35 percent of total home use.

By far, toilet flushing accounts for the largest percentage of water used in the bathroom.

The toilet is flushed about 20 times per day by the family of four, consuming more than five billion gallons of water in America each day. Wouldn't it be better spent adding aesthetic pleasure to our lives? Maximizing the enjoyment water can provide therefore means minimizing the amount we flush away with the nearly unconscious tug of the toilet's handle.

Showering and bathing – considered by many to be necessary for getting a good start on the day – consumes another 30 percent of the water used in the average household. A 10-minute shower or full bathtub for every American consumes three billion gallons of water every morning.

Another 11 percent of the water used in our homes flows from the bathroom faucet during grooming tasks such as teeth brushing, hand washing and shaving.

How To Get Water Smart

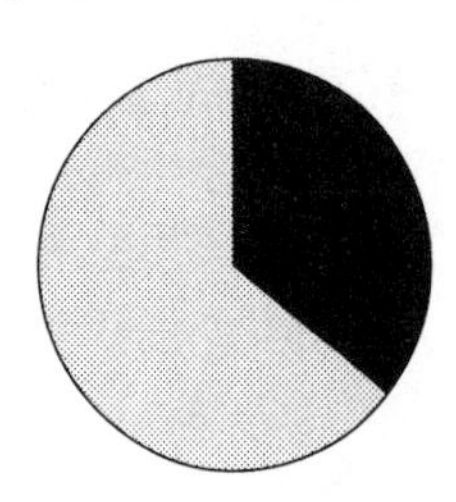

Bathroom 35% of home water use

Seventy-five percent of the water used inside the home flows from bathroom fixtures while we shower, bathe or groom.

When conserving water is important, the argument for installing new efficient plumbing technology holds plenty of water. Efficient new designs for the bathroom make savings automatic, leaving more of our household water for a little self-indulgence in the shower or bath. We humans find changing behavior and maintaining habits challenging, so improved technology can make bathroom conservation a sure success.

In addition to making conservation more automatic and realizing significant water savings, some designs will also delight the technologist in us with infrared sensors and instantaneous hot water.

Energy and water-saving goals must be examined in conjunction with the cost benefits and disadvantages of installing these new toilet, tub, faucet and shower technologies in the home. Further, the varying needs of large families and singles will dictate the best course of action to take in getting water smart in the bathroom.

Combining the installation of new technology with a bit of consciousness raising, detective work to eliminate leaks, and the cultivation of a few new habits by youngsters and oldsters alike can save hundreds of gallons of water every day.

Toilets

The toilet you flushed indiscriminately while you were growing up – with its Model A engineering – uses two or three times more water than the well-designed high-efficiency models of today. While the Model A is long gone, these monuments to a bygone era are still in operation in the bathrooms of most homes.

A quick analysis of family economics and water-saving goals will ultimately determine whether it makes sense to replace or "retrofit" these water-hungry commodes.

Early attempts at manufacturing low-flow toilets were rather dismal and resulted in clogged drains, repeated flushing by frustrated customers, and a generally bad reputation. Newer models have overcome these problems to a large extent with engineering and dedicated components that make better use of a small amount of water.

While most of us will change our flushing habits only the greatest degree of difficulty, best of all the new water-saving toilets require no environmental consciousness or extra work.

This section presents those products and practices that will help you tame the water closet beast.

Toilet Technology

Understanding the operating basics of the toilet and the function of water in the commode helps us make the best choice when considering the purchase of a low-capacity toilet – or just living with the one we already have.

To begin, understand that water serves not one but three purposes in the toilet. First, water is the medium in which waste is carried from the toilet bowl and out to the sewer. Second, water (propelled by gravity from the storage tank) provides the energy required for flushing to occur. And third, between flushings, water remaining in the toilet bowl and gooseneck trap seals off the sewer pipe, isolating the home from sewer gases.

More water is needed for flushing power than for either of the other purposes. In the typical toilet, with the tug of the handle a rubber stopper is pulled away from an opening in the bottom of the storage tank, allowing water from tank reservoir to travel at a high velocity through several bowl and rim openings. The new supply of rushing water then pushes the waste-ladden bowl water through the waste carry-out opening in the bowl and down the sewer pipe to freedom. Or, at least to the cesspool or waste treatment facility.

After the flush, the handle returns to its original position, the reservoir stopper falls back into place, and the storage tank refills. Commonly, some type of float rises with water until the reservoir fills to a specific level, causing the supply valve to shut off. In order to safeguard against overfilling the storage tank, an overflow tube protrudes slightly above the water line.

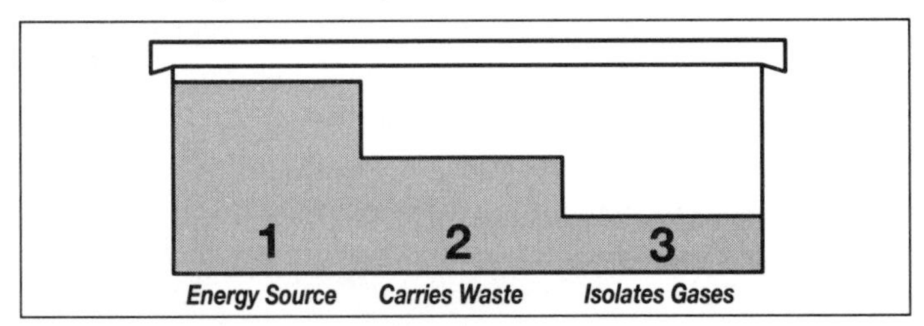

Water is used in the toilet in three ways: as an energy source; to carry waste from the toilet; and to isloate the home from sewer gases.

Why Replace?

Up until a few years ago, most household toilets used from five to seven gallons per flush to accomplish their mission. With the family of four tugging the handle 16 to 20 times per day, this meant 80 to 140 gallons of water would be consumed in a 24-hour period. New toilet designs offer the opportunity to save as much as 75 percent of this water, providing significant motivation for replacement.

One route to water savings is various retrofit methods for existing toilets that reduce the amount of water used to flush.

However, reducing water volume without upgrading other design factors can mean reduced toilet performance.

Another is the first modern water-saver toilet, the 3.5-gallon model. The original 3.5-gallon toilet was grudging manufactured in the late 1970s – sometimes with little change in actual toilet design – resulting in double flushes and clogged systems as unexpected standard features.

Growing pressure on manufacturers to develop the low-consumption toilet that uses only 1.6 gallons of water per flush meant vast changes in appliance design in order to preserve the necessary performance standards.

With city-sponsored rebates, water-efficient toilets can pay for themselves in 12 months or less

When to make the change? High water rates in some areas accelerate the payback of installing new technology. Conservatively, the new water-efficient toilets can pay for themselves in five to 12 years, unless your city offers a rebate for installing low-flow fixtures. This can lessen the payback time dramatically, perhaps to 12 months or less. Not bad, given that the average working life of a toilet is about 20 years.

Water Smart Savings: 80 to 140 gallons per day.

To Flush Or Not

Consider first whether there is need to flush the toilet every time you visit the bathroom. Your kids will love this concept because they never remember to flush the toilet anyway and they will adore repeating the annoying little rhyme, "If it's yellow, let it mellow/If it's brown, flush it down."

Of course, the amount of water saved varies with the number of flushes, but this tip only costs you a bit of discipline, tolerance and the price of a deodorizer. But if your family flushes only half as often, perhaps 10 times per day instead of 20, you'll save up to 70 gallons per day.

The good thing about not flushing is that it can save nearly as much water as a high-priced new toilet can. The bad thing is that the results can smell too. So if you make less-frequent flushing part of your water-saving repertoire, consider keeping the toilet lid closed and/or adding deodorizing agents to your storage tank and cleaning your toilet more frequently.

Water Smart Savings: 40 to 70 gallons per day.

Detrash Your Toilet

Let's continue our toilet tour with a bit of consciousness raising, such as the wasteful habit of using the toilet as a wastebasket. What other form of trash disposal uses five or more gallons of water to dispose of a tissue, a ball of cat hair or lint?

By installing a wastebasket in each bathroom, you can save water by a lot or a little. To find out how much, multiply the number of toilet flushes for trash disposal by the number of gallons of water your toilet holds.

Assume four family members eliminate this habit once per day. Four fewer flushes amounts to a water savings of at least 20 gallons per day – or 600 gallons per month – for the cost of an inexpensive waste can.

Water Smart Savings: 20 gallons per day.

Lock Up The Leaks

Surveys have shown that one out of five toilets leaks. When this happens, of course, water continuously flows down the drain while your water bill floats up. Two valves in your toilet most often account for unseen leaks, and these can waste hundreds of gallons per day – without a single flush of the toilet.

When the float valve wears out, the toilet tank is constantly filling while excess water escapes down the overflow tube. Alternately, if the stopper valve in the bottom of the tank fails, water constantly runs into the bowl. Either way it is gone for good.

When you do discover a leak, you'll need to adjust or replace the faulty valve with one of the following methods...or call in an expert to do so.

Bend An Arm

Diagnosing a poorly adjusted water inlet valve is a relatively simple task. Sometimes identified by a quiet gurgling or squeal, the leaky inlet valve can be an invisible but costly use of water in the bathroom.

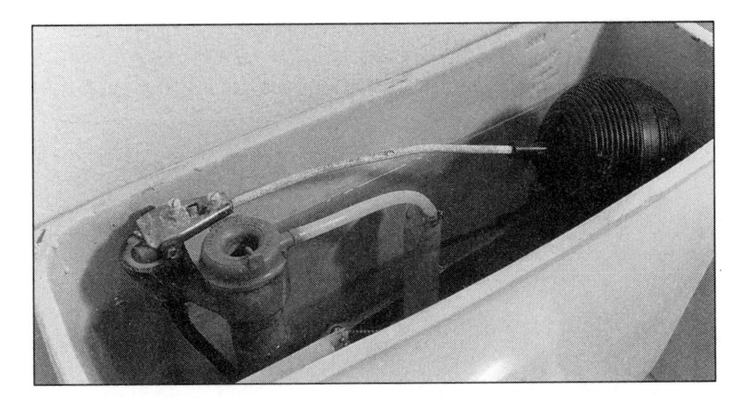

Repairing a poorly adjusted float or worn-out water inlet valve can save hundreds of gallons per day.

Start by lifting the cover off the toilet tank to see if water is high enough to escape down the overflow tube. This signifies either a poorly adjusted or worn inlet valve.

If your toilet uses a pivoting float and arm, you may be able to correct the problem by simply bending the float arm downward, causing the valve to shut off at a lower water level. By bending the arm still lower, you may be able to effectively reduce the tank capacity by a gallon or two. But remember, old-style toilets need plenty of water to function properly. Experiment and save.

Water Smart Savings: 40 gallons per day.

Listen Up

More insistent sounds of running water, or a loud continual squealing, may signify a worn-out water inlet valve which cannot be fully shut off, no matter how high the float rises.

If a worn inlet valve won't cooperate with the arm-bending technique, you may need to replace it. Replacement valve assemblies, complete with instructions, can be found at any good hardware store for about $6.

To replace the valve, first turn off the water main outside (doing so eliminates the chance of causing leaks at the toilet supply valve inside later on). Then simply remove the screws on the valve and lift out the float arm and valve plunger. Replace whatever parts the correct rebuild kit provides and reassemble. You're done, with enough money saved for dinner out.

Water Smart Savings: 40 gallons per day.

Color It Gone

A few drops of food coloring or dye placed in the toilet tank will quickly tell whether the stopper valve is leaking.

Here's a third toilet leak fix. A few drops of food coloring or dye placed in the tank will tell quickly enough whether the toilet's stopper valve is leaking. Color dye tablets are available at your hardware store or city water department, sometimes for free. Check the bowl after 10 minutes: If the water has turned color, the flush valve is leaking because of misalignment or wear.

Before you embark upon the task of replacement, make sure the existing stopper – either a ball or a flapper located at the bottom of the tank – is squarely aligned with its seat. If a modest adjustment here doesn't correct the problem, shut off the water supply and remove the stopper ball and the valve seat for cleaning with extra fine emery cloth.

Continued leaking probably means a trip to the hardware store for a replacement flush valve. Since toilets are pretty much universal creatures, a replacement valve should be readily available for about $6 or less.

Water Smart Savings: 40 gallons per day.

Retrofit Thy Commode

A variety of methods can be used to reduce the amount of water used per flush by your existing toilet. These work either by limiting the volume of the storage tank or by installing a variable flush assembly that cuts off the flow before all of the water rushes into the bowl.

Choose Your Flush

Several retrofit devices are available to increase the efficiency of the old family water hog while maintaining the energy available to perform the flush. Priced at about $20, these accessories allow selection of a full or partial flush for disposing of liquid or solid waste. While harnessing the full volume of water in the tank to create flushing velocity, the devices simply close the stopper sooner to save water. On partial flush, about two gallons is used to flush away urine. And when fully actuated, the tank spends itself entirely to clear solid waste from the bowl.

The award-winning Mini-Flush System can save three gallons per flush, or some 48 gallons per day for the average family.

One such device, the Mini-Flush System, has received a variety of design awards, including the 1990 Product of the Year award from the California Governor's office. The Mini-Flush System also comes highly recommended by EcoSource, a consumer's guide to environmentally sound technology.

Designed to work with an existing toilet, the Mini-Flush System can help prevent overflow in the event of a sewer back-up, and also reduces the load on septic systems.

When installed in a normal five- to seven-gallon toilet, the Mini-Flush System can save three gallons per flush. If set on "partial flush" for the 16 of the 20 times an average family tugs the handle each day, some 48 gallons can be saved, reducing toilet consumption by one-third.

The device installs easily without tools, and we figure it can pay for itself in water savings within three to six months, depending upon frequency of flushing and prevailing water rates.

The Mini-Flush System may be ordered for $19.95 through EcoSource, P.O. Box 1656, Sebastopol, CA 95473; 800-274-7040.

Water Smart Savings: 48 gallons per day.

Low-Tech Thrift

If you're crafty, you can slightly reduce the volume of water used in toilet flushing by installing one or two plastic quart bottles in the storage tank.

Start by placing a dozen small rocks inside a couple of empty plastic bottles. Then fill with water, replace the caps and insert the bottles into the toilet tank. Make sure you position the containers so they won't foul any operating parts of the tank. One plastic bottle will generally fit a 3.5-gallon tank, and two should fit into a five-gallon unit.

While saving less than a gallon per flush may seem relatively insignificant, the savings add up over days, weeks and months. Assuming that a family of four modified its five-gallon toilet to emulate a 4.5-gallon unit and flushed the unit 16 times per day, water savings would amount to eight gallons per day – or more than 240 gallons per month.

The only pitfall here is that some valve acutating systems may not leave room for this kind of impromptu volume reducer. Also, reducing the amount of water stored in the tank reduces the amount of energy available for full performance of your toilet. But if it works, you win.

Water Smart Savings: 240 gallons per month.

Dam That Toilet

Toilets with capacities of five to seven gallons are good candidates for a newer type of volume restrictor, toilet dams. These flat dividers are placed in the bottom of the toilet tank and operate like baffles to hold back from one to two gallons of water per flush.

To install, first turn off the water supply and remove the top of the toilet tank. Flush the toilet just once to remove water from the tank, then place one toilet dam on each side of the drain valve, making sure the dams' rubber edges press against the bottom and sides of the tank. It is important to assure that each toilet dam is properly installed, because if the dam works free, it can interfere with toilet mechanisms, resulting in more waste than savings.

When ready, turn the water valve back on and operate the toilet again to observe the flushing action. If the flush isn't adequate, move the dams farther away from the flush valve or remove the dam closest to the overflow tube.

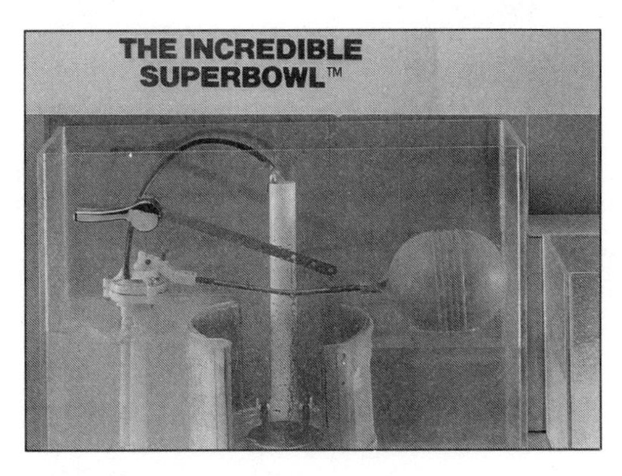

Toilet dams hold back one to two gallons per flush, saving up to 32 gallons per day or 960 gallons per month.

One popular toilet dam, The Incredible Super Bowl Saver, is available at many hardware and plumbing stores for about $7. In a five-gallon toilet, the dam can save up to two gallons per flush, or 32 or more gallons per day for an average household. At the end of a month, you could save as much as 960 gallons.

One last suggestion about retrofitting: Don't place a brick in the storage tank to displace water. Bricks will eventually disintegrate, resulting in a new set of annoying – and expensive – problems.

Water Smart Savings: 960 gallons per month.

Biting The Low-Flow Bullet

To achieve the ultimate in water savings while maintaining performance, the ultimate option is to replace your old dinosaur commode with one of the many ultra low-flow toilets now available.

All across America, manufacturers have gotten smart to the concept of good-performing low-flow toilets. Finally catching up with the need for water awareness, engineers have penned new lines of 1.5- to 1.6-gallon toilets that perform well while still appealing aesthetically to even the most discriminating buyer.

The cost of these new ultra low-flow toilets ranges from under $100 to more than $300, plus of course installation costs. Deciding whether to replace the family toilet involves a number of factors, including the size of your family, water and sewer rates in your area, purchase and installation costs and the number of bathrooms in your home.

The American Standard New Cadet Aquameter offers 1.5-gallon-economy with traditional toilet styling.

Some cities and water districts offer rebate programs as an incentive for replacing old hardware, and the momentum to provide this encouragement seems to be growing at the state level. Rebates of $50 to $80 may be available from these districts, so by all means first check with your local water purveyor for information.

In one California community of 100,000, some 10,000 such rebates were handed out in a single year. What did the city dump do with all the old toilets? They were crushed and recycled as part of the road base on city streets.

Some cities and water districts offer rebate programs as an incentive for replacing old hardware, and the momentum seems to be growing at the state level.

The water-savings argument for replacement is very strong. Consider the enormous annual water savings produced by replacing a five- to seven-gallon toilet that consumes 80 to 140 gallons per day with a 1.5-gallon model that consumes about 30 gallons per day given the same frequency of flushing. Depending on water rates in your area, the low-consumption commode can pay for itself in just a few years. At the least, it will free up water for other uses.

Informational bulletins containing listings of toilets, urinals and flushmeter valves that meet water conservation standards of the American National Standards Institute are available in many states. For a specific list, contact the Division of Codes and Standards at the Department of Housing and Community Development in your area. And don't forget to consult publications such as *Consumer Reports, Consumer Guide* and *Practical Homeowner* when considering specific new products.

Finally, the National Kitchen & Bath Association is an international organization that sets professional standards for related businesses. You can obtain a current directory of accredited members by contacting the NKBA at 124 Main St., Hackettstown, NJ 07840; 201-852-0033.

Water Smart Savings: 50 to 110 gallons per day.

In Search Of The Perfect Flush

When considering whether to retrofit your existing toilet or replace it with a new water-saving model, you need to consider the few trade-offs involved.

Common sense tells us that no matter how much water is used in a toilet's operation, waste should be carried away in one flush and the bowl should appear rinsed and free of unsightly "skid marks." Further, a low-consumption toilet should ideally not increase the frequency of clogging.

Several design variables play into the efficient operation of the toilet. First is the trap seal, the amount of water left in the bowl at the end of the flush cycle. The trap seal prevents noxious gases from filtering into the home from the sewer pipe. At least 2.5 inches of water is needed in the bowl to accomplish this task.

The water spot is the surface area of the water standing in the bowl for the purpose of catching deposits. The traditional toilet has a water spot of between 8 x 10 inches and 10 x 12 inches. This surface area plays a large role in the amount of bowl staining that occurs. While our European neighbors grow up with small water spots and simply keep bowl brushes handy, American manufacturers have learned that the domestic consumer has little patience for such practices.

Another design feature to look for is an effective "rim wash," the rinsing action that occurs on the walls of the bowl during flushing. The effectiveness of the appliance's rim wash depends on the size and shape of the openings in the rim and the direction of water flow.

Varying toilet designs and technologies determine the mode of waste carryout–that is, the effectiveness of waste movement out of the bowl. Likewise, the design and placement of the trapway or waste hole also becomes a factor in the efficiency of the toilet.

WCI's pressurized flush system harnesses household water pressure to force a small amount of water out of the tank at high velocity.

Rising Stars

Several design innovations are employed in the newest ultra low-flow toilets to reduce water consumption while preserving flush performance. Most interesting, a pressurized flush system patented by Water Control International harnesses the water pressure within the supply system to compress air in a storage tank. When the toilet is flushed, this pressurized air literally "pushes" a small amount of water out of the tank at high velocity.

The WCI Flushmate

The Flushmate, Water Control International's patent name for the technology, retains the typical 8 x 10-inch water spot, reducing unbecoming bowl stains. Combining other critical dimensions that WCI says are superior to 3.5-gallon European designs, the Flushmate is said to be able to clean the bowl of waste in a single flush with 1.5 gallons of water.

Today several well-known toilet manufacturers employ WCI's Flushmate engineering. *Consumer Reports* rated this technology second only to a well-designed traditional gravity-flush system. On the down side, *CR* noted, WCI's pressurized flush is somewhat noisier than other toilet designs (but only for a few moments), and some elders and youngsters may have difficulty pushing the flush button located atop the toilet tank.

Flushmate-equipped toilets are offered by major makers American Standard, Crane, Gerber and others. As styles, colors and details vary, it's best to compare before choosing. And with sticker prices crowding the $300 figure, it's a good idea to shop for value as well.

Complete information on the Flushmate system may be obtained from Water Control International, 51155 Grand River Ave., Wixom, MI 48393; 313-349-5300.

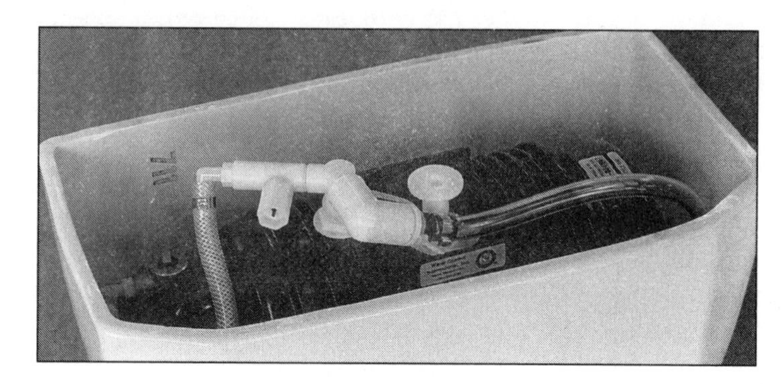

Unique Flushmate design is said to be able to clean the bowl of waste in a single 1.5-gallon flush.

Kohler

Other new toilet designs, including Kohler's Wellworth Lite, tackle the low-flow issue a bit differently. While using just 1.5 gallons of water per flush, the Wellworth Lite employs a twin-valve setup designed to emulate the thorough flushing action of a toilet with much more water to spend.

Kohler's Wellworth Lite uses twin flush valves to emulate the flushing action of a toilet with much more water to spend.

With the Kohler system, a pilot valve works in conjunction with a companion flush valve to initiate – and then follow through – with a strong flushing action without the use of compressed air. Advantages of this system include consistent performance without regard to available water pressure.

The Wellworth Lite received the top *Consumer Reports* performance rating, and is claimed by the manufacturer to exceed all ANSI standards for standard 3.5-gallon toilets. This particular Kohler toilet sells for about $240.

Universal-Rundle

You might think of the low-flow Universal-Rundle toilet as the Chevrolet of commodes. The car folks in Detroit may not like the analogy, but what it means is that Universal-Rundle Atlas model is designed to be gimmick-free, with more form and function than its fancy Euro-style competitors. Those are some of Universal-Rundle's words. Also promised – in fact, guaranteed – is that the Atlas will flush problem free under normal conditions.

Touting refined traditional technology as the answer to the 1.5-gallon question, the $109 Atlas uses a conventional ballcock and flush valve combined with a "high performance" bowl design for dependable as well as water-efficient flushing.

The Universal-Rundle Atlas, using refined conventional technology, is an inexpensive choice for family bathrooms.

A complimentary 3.5-gallon model is also available for households with a little more water to spare. Contact your plumbing store or Universal-Rundle Corp., 303 North St., New Castle, PA 16103; 412-658-6631.

No-Water Toilets

For extremely water scarce areas or home sites where neither sewers nor cesspools are allowed, waterless incinerator toilets are a viable option.

Simply defined, incinerator models change waste to ash electrically. These models use no water and need no plumbing, no septic systems...although they do require electricity. Best of all, they are essentially non-polluting.

Incinolet produces a range of incinerator toilets including the Carefree Deluxe model, which was designed for use in homes and cottages. Prices range upwards from $1,295. Contact Research Products, 2639 Andjon, Dallas, TX 75220; 800-527-5551.

Water Smart Savings: 80 to 140 gallons per day.

Installation Tips

Look into your toilet's "footprint" before you leap into the low-flow camp. A new toilet with a smaller footprint may require floor repairs or even replacement.

When the time comes to replace a toilet, be sure to consider the commode's "footprint" or base shape. A different base shape may affect the appearance of your bathroom floor after replacement. Many of the low-flush toilets have a smaller footprint, requiring that the floor covering be patched, modified or replaced.

Additional costs of replacing your toilet will include a new wax gasket (less than $2), new mounting bolts and a new toilet seat (starting at less than $20) as well as the cost of professional installation, if desired. Figure on about $40 labor per toilet.

Bathtubs

If a sailboat is defined as a hole in the water into which you pour money, a bathtub is defined as a hole in the bathroom into which you pour water. Thus, the ageless tub is capable of consuming more water – and money – at one time than any other indoor appliance you possess.

Soaking up to your neck in bubbly hot water can use 40, 60, 80 or more gallons of wet stuff in one gulp, accounting for 10 percent of the water used in the home.

These days we're left with quite a quandry. While mental health professionals may advocate a relaxing warm bath to ease the day's tensions, water conservation experts advise against this type of water use altogether. So what's a water-conscious, stressed-out person to do?

Altmans Mini-Riveria provides total luxury with optional whirlpool, fills with only 40 gallons.

Bathtub manufacturers and the designers of bathroom whirlpools and spas have come up with precious few technological advances that make bathing guilt-free. For instance, low-flow faucets make little sense when the end result is to fill a bathtub to capacity.

The advent of personal Jacuzzis and Whirlpools, not to mention double-occupancy tubs, can also mean the use of a hundred or more gallons of water at one time. Obviously, the move toward bathroom luxury makes water-smart choices all the more important.

Do More With Less

The average household bathtub can hold 40 to 60 gallons when filled to capacity. But there is no law against turning off the bathtub faucet before the tub fills completely.

Bathing advocates will be pleased to note that filling a standard bathtub only a third or half full can save 20 or more gallons compared to drawing a full bath. In fact, under certain circumstances, bathing in a partially filled tub actually uses less water than all but the shortest of showers.

Bathing in a partially filled tub uses less water than all but the shortest of showers.

To find out the facts in your house, you can run this simple test: On the first day, take a shower for the normal length of time with the bathtub drain closed. At the end of your shower, mark how high the water is in the tub with a grease pencil or a dab of lipstick. If you can bathe by using less water, you'll save water by taking baths instead of showers. At the very least, filling the tub only half full will save 50 percent over taking full baths.

Water Smart Savings: 20 gallons per bath.

Explore Your Options

The relaxation value of a warm bubble bath is tough to dispute. But you may be able to find other modes of relaxation in your lifestyle to replace bathing as a daily meditation. For example, sipping a warm cup of cocoa and reading a good book while curled up on the couch may serve just as nicely to unwind from a stress-filled day.

An evening walk by yourself or with loved ones is another powerful relaxation tool, one that provides aerobic benefits. And how about yoga or massage? Whether provided by a professional, a family member or friend, massage is a wonderful source of relaxation that identifies and gets rid of tensions as well as – or better than – the proverbial warm bath.

Net Savings: 40 gallons per day per person.

Stop It Up!

Put the stopper in the tub drain before you turn on the spigot. By the time the tub fills to the desired level, the small amount of cool water that first entered the tub will be nicely brought up to temperature by the warmer water that follows.

Household water recirculation systems and point-of-use water heaters are other, more elaborate systems discussed elsewhere in this book. However, for bathing, the practice of stopping up the tub before water is turned on generally eliminates the need to employ this energy-consuming, expensive technology to save water. If you are remodeling or building a new bathroom, you can explore the benefits and costs of these systems for the tub, shower, and other areas of your home.

Water Smart Savings: Five gallons per bath.

Take Faucet Control

KWC and some other plumbing manufacturers offer temperature and flow-control devices that allow you to set the faucet flow rate you want while maintaining water temperature within two degrees. These "stabilized" systems can reduce the amount of water flowing through the faucet by some 20 to 50 percent – good news for adults who find passing water-saving philosophy along to kids a challenge. Better news still is that the faucets can eliminate the danger of scalding.

Even with flow-restricted faucets, the bottom line in bathtub use is still tub capacity and good habits. Mastering these will allow you to bathe without guilt.

For information on the KWCtherm system shown below, contact Western States Manufacturing Corp., 1559 Sunland, Costa Mesa, CA 92626; 714-557-1933.

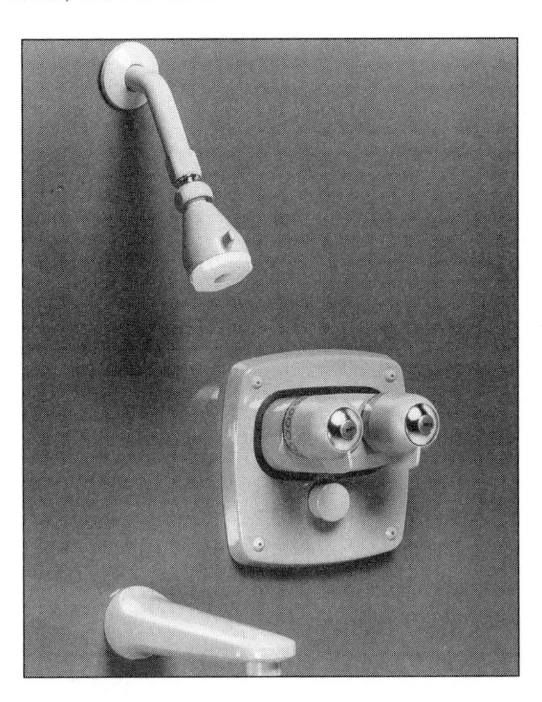

KWC offers a pressure balance control device that allows water flow rate and temperature limits to be preset, thus reducing water use – and scalding.

Bathe With A Buddy

Kids and adults alike can take great enjoyment in bathing with a partner. Doubling up small children is an obvious water saver. Plus, adults may enjoy certain romantic benefits from bathing with a significant other.

A dash of Mr. Bubble and a tub full of toys will likely quiet any complaints about having to share one's bath water, even among some adults we know. Plus, since two bodies displace twice the water as does one, you'll use less water to fill the tub in the first place.

A variation on the theme for multi-child households is the concept of successively bathing children in the one tub full of water...assembly-line style. Drawbacks for those low on the list may include cold, crummy water. Don't worry, you can drain half of the water, refill with hot, and still come out ahead in the volume game.

Fixing a flowing pinhole leak can save 170 gallons per day or over 5,000 gallons of water per month.

When considering multi-child bathing, it's important to impress upon kids that urinating in the tub is to be avoided. You can probably find your own words to use. So require your child to visit the toilet before stepping in that nice warm tub. If your child gets that far-away look in his eye upon entering the tub, you might reconsider sharing that water with others.

Water Smart Savings: 40 gallons per day.

Nailing Down Leaks

A dripping bathtub faucet can waste hundreds or even thousands of gallons of water in a month. At one drip per second, a leak accounts for about 200 gallons of water per month. Two drips per second will send 400 or more gallons down the drain.

When a drip turns into a real leak, the numbers turn downright scary. A flowing "pinhole" leak, for example, is something few of us could ignore. In full bloom it will cost you 170 gallons per day or over 5,000 gallons per month. Ignore it if you want, but that's the cost. No useful purpose has been met other than rusting your tub and boosting your water bill into the stratosphere.

Another water waster in the tub is a leaking stopper that allows water to escape from the tub during bathing. Such a leak makes adding more water to the tub tempting and also makes it possible to take a bath using even more water than the tub has capacity. Thus, check your bathtub faucet and stopper for leaks several times a year.

To fix a tub faucet leak, shut off the household's main water supply, unscrew the offending faucet handle and remove the "escutcheon," the ornamental chromed piece that covers the faucet assembly. Underneath is the faucet valve, which will need to be disassembled and rebuilt by means of the appropriate kit.

Adjustment or replacement of the stopper mechanism can be accomplished by first finding the access hole that allows you to reach the lever mechanism. That access point is found in the room, closet or wall directly behind the bathtub. Adjust the linkage to improve operation, and replace the stopper seal if worn.

Water Smart Savings: Up to 5,000 gallons per month.

On Larger Tubs

When remodeling bathrooms, these days many families opt to replace the traditional five-foot tub with a larger double occupancy tub, Whirlpool or Jacuzzi. According to the NKBA, an estimated eight million bathrooms were built or remodeled last year, many sporting these new luxury fixtures. Of all the new tub fixtures sold in 1990, some 18.6 percent were of the Whirlpool/ Jacuzzi variety.

Among the luxury elite is American Standard's Sensorium, which can be equipped with a computerized whirlpool and environment control and communications system. The Sensorium can also be preset to fill and maintain a specified temperature as far as 24 hours in advance, while an available remote control system even sets light and music levels. Most importantly, the whirlpool system allows you to set water depth at low, medium and high (61, 69 and 79 gallons). The cost is exotic as well – Sensoriums are priced from $5,243 to over $25,000.

More information on the Sensorium system can be obtained by calling American Standard at 800-821-7700.

American Standard's Sensorium, a luxury water experience, automatically creates a complete environment for bathing enjoyment – with low, medium or high water level.

Please note that the capacity of most indoor spas is generally quite a bit greater than the traditional family tub. Before purchasing, it probably makes sense to consider a traditionally sized bath equipped with a whirlpool system, rather than a voluminous two-person model. Most two-person units we're aware of require 80 to 100 gallons of water to fill. Also, choose a model that can work with less than a full tub of water. Most systems need a foot or more of water to operate correctly.

On Reusing Tub Water

As the use of grey water becomes more common, health officials are beginning to provide guidelines for its use.

Health factors in reusing household water – also known as grey water – must be carefully considered. Certainly, water captured in a bucket while the water is warming is not considered grey water and can safely be used for any household purpose. But when either children or adults are bathing, chances are good that water will be contaminated with various bacteria. Hence, bath water should never be used for watering vegetable gardens or for any other purpose that may involve human contact.

Slightly used bath water may be safe for outside shrubs and plants as long as no one will be coming into direct contact with areas watered. The reuse of bathtub water should be avoided when illness is present in the household.

As the use of grey water around the home becomes more common, public health officials are beginning to provide guidelines for its use. Contact your local state Department of Health Services or a county health officer for precise information.

Showers

The invigorating shower is surely the most popular form of bathing. Showering accounts for 20 percent of all the indoor water used in a home, so the need for installation of new technologies and implementation of water-saving habits play key roles in the shower stall as well as bathtub.

Luckily, new showerhead technologies make water conservation in the shower easy – with little loss of shower effectiveness or enjoyment. When these products are combined with small changes in showering procedure, more than 80 percent of the water that used to go down the drain during showering can be reassigned to other purposes.

This section is devoted to informing you about these new devices and habits that maintain showering as a pleasant experience – while still saving water.

Don't Just Stand There

We may as well face the inevitable: One of the simplest ways to save water in the shower is to spend less time under that dreamy flow. We've found that the average 10-minute shower can be easily cut in half, thus saving half of the water normally used in the shower – no matter what kind of showerhead you have.

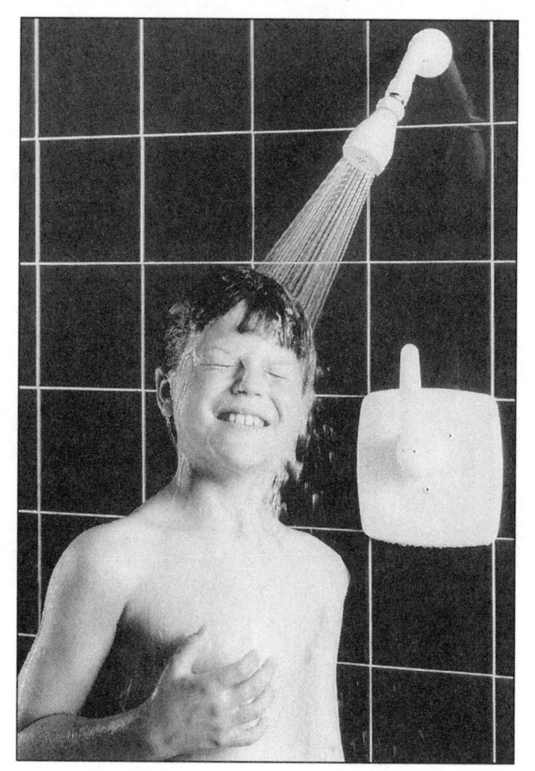

Reducing the time spent showering from 10 to five minutes can save up to 48 gallons per person each day.

Showering for five instead of 10 minutes while using a low-flow 2.4 gallon per minute showerhead instead of a standard head (3.5 to 6.0 GPM) means you use 12 gallons of water instead of 35 to 60 gallons, a 65- to 80-percent savings. Over the course of a year, this amounts to between 8,400 and 17,500 gallons saved per person showering daily in your household.

Water Smart Savings: 23 to 48 gallons per shower.

Installing A Flow Restrictor

A simple round plastic washer containing a small hole or multiple holes – commonly called a flow restricter – can be installed in your existing showerhead to reduce water use significantly at little or no expense. Most water departments offer these restricters free of charge and installation is simple.

Leaks from the shower faucet should be extended the same welcome in your home as ants in the pantry.

To install the washer, unscrew the showerhead by hand or with a wrench, being sure to cover the showerhead with cloth or tape before using the wrench. The flow-restrictor washer should slip easily into the back of the showerhead nozzle. Reattach the showerhead snugly by hand or wrench.

The biggest problem with this makeshift technique is that as well as reducing water use, the resulting drippy stream also can reduce the quality of the shower.

Water Smart Savings: 30 gallons per day.

More On Leaks

Leaks from the shower faucet should be extended the same welcome in your home as ants in the pantry. For shower leaks, like those in the tub, can account for far too much of your home's water consumption to be tolerated.

Fortunately, the same faucet you fixed to banish tub faucet leaks may also control the shower. So if you have a leaky shower that needs fixing, follow the directions found in the bathtub section of this chapter.

Water Smart Savings: Up to 1,000 gallons per month.

Replacing The Water Guzzler

Most showerheads installed prior to 1980 are flow rated at six gallons per minute, although we have found that their actual output may range from 3.5 to six gallons depending on household water pressure. Under high-pressure conditions, however, one of these devices will indeed flow 60 gallons during a 10-minute shower.

Legislation in many western states now prohibits the sale of these guzzlers, requiring instead a maximum flow rate of 2.5 to three gallons per minute. With the installation of these low-flow showerheads, available at a costs ranging from free to nearly $100, you can reduce water used in the shower by up to 60 percent, while still taking that luxurious 10-minute time out.

The new showerheads employ a narrow spray area so that less water misses you and may also add air to create a turbulent flow. Regardless, the ones we've tried still work well enough to provide an enjoyable shower.

We've discovered that the sheer volume of water delivered doesn't determine shower satisfaction. More important is the quality of the spray including its breadth and consistency. And surprisingly, since the accumulation of mineral deposits can in time reduce showerhead performance, you may well find that replacing the old water waster with a low-flow head can actually mean improved rather than decreased performance.

Low-flow showerheads are sometimes available free of charge from water city departments in water-stressed areas. One such unit by Niagra is flow-rated at 2.4 gallons per minute. It has the added benefit of an easy-use shutoff pin that allows halting water flow (except for a few drips) while lathering up. If your water department can't help, buy one for $6 at your local plumbing store. You'll be glad you did.

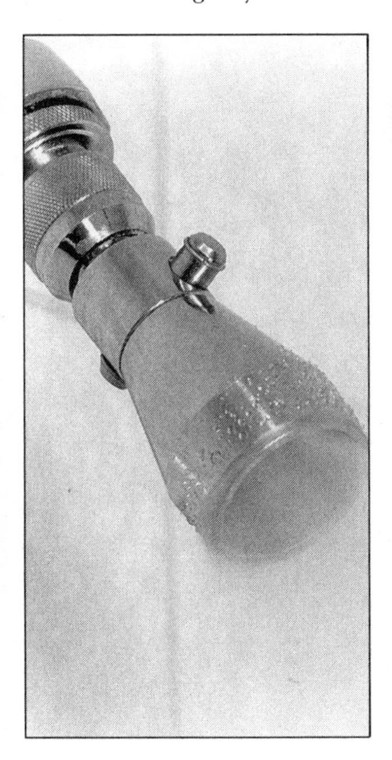

The Niagra 2.4 GPM showerhead with shutoff retails for $6, but this one was obtained free from a local water department. It provides good spray quality and a 60-percent water savings.

To determine if your nozzle is a guzzler that should be replaced, check its flow by placing a half-gallon container directly under the showerhead and then turning on the faucet full blast. If the container fills in much less than 10 seconds, you've got a guzzler showerhead that can be replaced with a low-flow device.

You can obtain a directory of certified showerheads from most state Energy Commissions or state water departments, and research through the popular consumer publications will provide information about the best – and worst – currently on the market.

Water Smart Savings: 36 gallons per shower.

Top Performers

Several showerheads have come highly recommended by *Consumer Reports*, which considers both performance and price in its rating. Receiving fine reviews are the Sears Energy Saving Shower Head model 20170 ($6), Thermo Saver DynaJet ($6), and The Incredible Head ($7) from Resources Conservation. All these models worked best with moderate and high water pressures found in most homes.

The Incredible Head makes* Consumer Reports' *list of recommended low-flow showerheads, is available from* Resources Conservation. *Call them at 203-964-0600.

Best Of The Best?

If you are remodeling or building a new shower unit, you'll of course look for a shower faucet system that provides maximum water efficiency as well as other features. To suit, KWC Faucets offers the KWCdomo pressure balance shower system, which offers both a water flow control and a constant temperature feature that eliminates the chance of scalding. The result is consistent water savings in the shower, with greater comfort and peace of mind.

Such control doesn't come cheap. The KWCdomo in chrome runs $421, and color-keyed valves cost $477. KWC products are distributed by Western States Manufacturing Corp., 1559 Sunland, Costa Mesa, CA 92626; 714-557-1933.

KWC offers the KWCdomo pressure balancing shower mixer that can be preset to both restrict flow rate and control water temperature – at a price.

Replacing The Showerhead

It's easy to replace your showerhead with a new water-efficient unit. Use your hands or a wrench padded with cloth to unscrew the old showerhead. Clean the threads of the water outlet pipe, add teflon tape to the threads, and attach the new showerhead.

If the new model won't fit on your existing shower neck, you may need an adapter kit to match the water delivery pipe to the new head. These are available at plumbing and hardware stores for a few dollars. The truly water-smart buyer will bring the old showerhead along on the shopping trip for the new head, ensuring that the new product will fit.

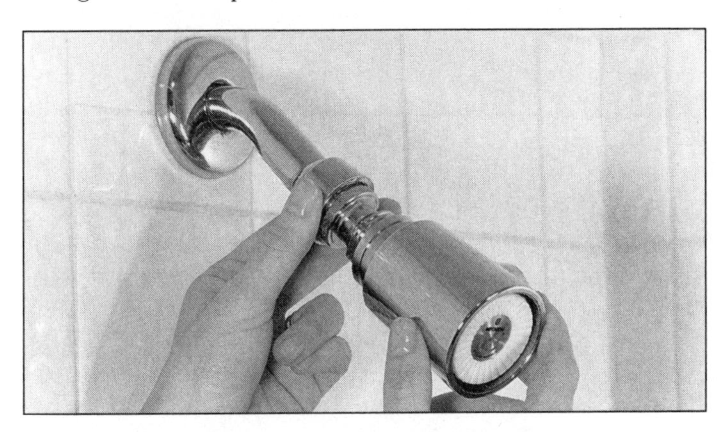

Replacing your existing showerhead with a low-flow model such as this 2.5 GPM Moen 3905 can be accomplished in just a few minutes.

Why Even Shower?

One mayor we know of recently announced to her constituency her own contribution to water conservation – showering or bathing every other day. Predictably, the local media had a field day with this bit of news, and rumor had it that there was a scramble to reschedule meetings with the Mayor to coincide with her Shower Days.

While the every-other-day method may work for the occasional mayor, parents could well have difficulty co-habitating with active children using this method. And if your own work is extremely strenuous or dirty, you may find your circle of friends growing smaller.

Still, it's a thought.

Water Smart Savings: 210 gallons per person per week.

Shut Your Tap

Turn off the shower while shampooing or soaping up to earn extra water smart points. Many of the low-flow showerheads feature a temporary shut-off button that maintains constant temperature and pressure while you temporarily cut off the supply.

If you try this just once, you'll find that the shower will be off at least 50 percent of the time you spend in the shower. Corresponding shower water savings, and a sense of accomplishment, will result.

This practice may be tough to maintain on particularly cold mornings, or if you've just windsurfed around Nantucket in February. Have a nice fluffy towel ready.

In case you've already got a low-flow showerhead – but no shutoff button – look for Omni Products' model 711 On/Off Water Shutoff Valve. It costs about $12 at your local plumbing store.

Water Smart Savings: 15 gallons per shower.

Shave At The Sink

Even at a low-flow 2.5 gallons per minute, shaving in the shower can consume a whopping 12 gallons of water.

It shouldn't take a Mensa IQ club member to figure out that shaving in the shower can use five or more minutes' worth of water – at shower flow rates.

Even at a low-flow 2.5 gallons per minute, this practice consumes a whopping 12 gallons of water. It's best to shave after showering, running the tap only as needed or partially filling the basin with warm water. As a bonus, your skin will already be soft and warm, and shaving will be easier.

Water Smart Savings: 12 gallons per shower.

Capturing The Cold

If you are like most people, there is no way you'll step into a shower stall until you are sure the water is sufficiently warm to the touch. No one will blame you for that. But when you turn on the shower, at least start by turning on only the hot water valve until the water heats up. Mix in the cold to adjust the temperature only after water temperature has risen.

And if you need to adjust the flow or temperature of the water at any point during the shower, subtract the hot first, then the cold to realize both energy and water savings.

Water Smart Savings: 30 to 60 gallons per person per month.

Rub-A-Dub-Dub

Make a habit of using a bath brush or wash cloth to remove stubborn dirt from your body instead of depending on the force of the shower water to do the work.

Wash cloths and bath brushes can be combined with turning off the shower while lathering.

Water Smart Savings: Five gallons per shower.

Instant Hot

Installing a point-of-use water heater is an innovative way to make sure hot water gets to the showerhead – and other faucets – right now. Most often used in commercial and recreational settings such as motels, schools and recreational vehicles, the instant-hot technology can be harnessed for home use too.

Typically installed under a counter or left free standing, point-of-use heaters provide hot water without waste. This can be especially important in homes or apartments where the hot water heater is seemingly miles from the faucet.

There are a number of brands on the market. One, the French-built tankless Aquastar, is fueled by propane or natural gas and should operate more efficiently than electric models. Although attractively styled, the Aquastar still may face a tough climb before it's accepted into the American bathroom. Prices start at $450 from Real Goods Trading Corp., 966 Mazzoni St., Ukiah, CA 95482; 707-468-0301.

For severe hot-water shortages, a secondary heater can provide instant hot water...along with water savings.

The downside of all secondary water heaters includes purchase price, complexity, and increased energy costs. More information on point-of-use water heaters can be found in the Sinks & Faucets section of this chapter. Please also refer to the Auto & Garage chapter for information on household water recirculation systems, which can provide hot water instantly throughout the household.

Water Smart Savings: 30 to 60 gallons per person per month.

Sinks & Faucets

Many types of equipment have been developed to aid water conservation at the bathroom tap, where about 11 percent of water is used in the home on a daily basis.

The main source of water waste lies in the practice of turning the faucet on and leaving it on while brushing teeth, washing hands, combing hair, shaving or perhaps just daydreaming.

New water-conserving faucets, flow restrictors and aerators, and state-of-the-art automatic infrared faucets, may be combined with some important changes in sink habits to reduce water loss.

Use It Up Or Shut It Down

As creatures of habit, many of us are accustomed to allowing the water tap to flow while we use the sink. Simply put, we value our attention and labor more highly that the water that is used. Of course, water smart rule number one for the bathroom sink is to run the faucet only when you need it.

Using the sink stopper will enable you to capture water for extended rather than momentary use.

Making frequent use of your sink stopper will enable you to capture water for extended use rather than the momentary use you get from each droplet of flowing water. Best of all, engaging the sink stopper scarcely takes any more effort than turning on a tap, and certainly it takes less than constantly turning the tap on and off. Just make sure that the stopper is adjusted properly for a good seal.

Water Smart Savings: 30 gallons per day.

On Brushing Teeth

Running the tap while brushing your teeth consumes three to 10 gallons of water that never touches your lips or gums. We seem to find something comforting about hearing the water splash in the sink while we polish our pearly whites. In a water-conscious household, it's a habit worth overcoming.

Try turning on the faucet only long enough to thoroughly wet your toothbrush and capture water in a glass. Use the water in the glass to rinse instead of rinsing your brush repeatedly under the water flow. This practice consumes scarcely more than a cup of water. Plus it works great.

Water Smart Savings: Two to nine gallons per person per day.

On Shaving

When it's time to eliminate that scratchy growth on your face, put the stopper in the sink and fill the basin with hot water before you shave. You'll save the five to 10 gallons of water normally lost while shaving with the faucet running for five minutes.

The same practice can be employed for washing hands. Rinse your hands quickly under the faucet with the stopper in place and turn off the water while soaping. Then rinse your soapy hands in the water remaining in the basin.

Three or more gallons of water are used when washing hands the old-fashioned way. The new practice will use a gallon or less.

Water Smart Savings: Three to 10 gallons per person per day.

The Low-Flow Way

Before purchasing a new flow restrictor, remove the old one and check whether it features male or female (outside or inside) threads. If your faucet spout is not removable, it's likely quite old, and may need replacement or at least a special adapter before you can enjoy low-flow savings.

Flow restrictors are readily available at any hardware store, plumbing shop or home improvement center, and sometimes for

Installing a spray tap, restrictor or aerator is easy and can save nearly five gallons per minute.

free at city water departments. Models range from 0.5 gallon per minute up to 2.75 GPM. How much water they will actually flow depends on your home's water pressure. For bathroom purposes a 1.5 GPM aerator will be quite sufficient, while providing a tremendous water savings.

Water Smart Savings: Up to five gallons per minute.

Repairing Leaks

As with shower and tub faucet leaks, a fast sink drip can waste vast quantities of water. In addition, water lost due to a poorly seated stopper can add up by requiring you to refill the basin.

To repair a leaky faucet, first shut off the water supply, then use the appropriate wrench, screwdriver or allen wrench to remove the faucet handle. Remove the stem or spindle and check the washers you find inside for wear. Also check the washer seat for smoothness. Most faucets are designed to allow replacement of O-rings or seals for long-term maintenance. You should be able to find these individually or in a kit, so take the stem with you when you visit the plumbing store.

To adjust the sink's stopper, look for the actuator assembly underneath your sink. You'll find one or two clamps or thumb screws that allow you to adjust the mechanism. At the same time, check the wear of the stopper seal by releasing the stopper from its assembly and pulling it out of the sink. Replace the seal if necessary.

Water Smart Savings: Up to 1,000 gallons per month.

Faucet Choices

In years past the standard faucet flowed at a rate of five gallons per minute. New faucets flow at a rate of 2.75 GPM or much less, providing a substantial argument for replacing the old water waster.

Most states provide a listing of certified and approved low-flow faucets. To obtain such a list, begin by contacting your local water department, or try the National Kitchen & Bath Association at 800-367-6522 for a list of member dealers and manufacturers. Your local plumbing retailer should be able to help, too.

Before you buy you'll need to know the size of the faucet holes in the sink. You will also need to measure the distance from the center of each mounting hole to the spigot. Most faucets are built for four-inch centers.

Omni flow control devices can reduce lavoratory water flow by 45 to 90 percent while producing a clear stream.

Omni Products builds and sells lavoratory flow control devices that offer a claimed 45 to 90 percent water savings over unrestricted faucets. These units are said to provide a constant flow rate through the faucet no matter whether the household water pressure is 40 or 100 PSI. And because they also do not aerate the water stream, the Omni restrictors are said to provide a satisfying stream of clear water and reduced splashing. Costs range from $7 to $8.

Omni and other manufacturers offer flow restrictors that feature an on/off button located on the faucet spout that can temporarily cut off the flow of water without need to readjust controls. This technology helps you use just the amount of water that's needed.

Contact your retailer or Omni Products, 21011 So. Figueroa St., Carson, CA 90745; 800-447-4962.

At the high-tech end, Water Facets offers its new Contempra line of faucets that feature infrared sensors that detect any object – a hand, a toothbrush or a razor – placed under the faucet. When the sensor determines an object is present, a microprocessor opens the faucet's solenoid valve and allows water to flow. When the object is removed, the valve closes and the flow stops. Temperature may be preadjusted with a single control knob.

The system is equipped with a continuous flow button to override the sensor when needed, such as when filling a basin. The Auto Water Faucets come in four- and eight-inch faucet centers to fit most sinks. This type of faucet may be installed with a point-of-use heater or a home water-recirculation system for maximum effect.

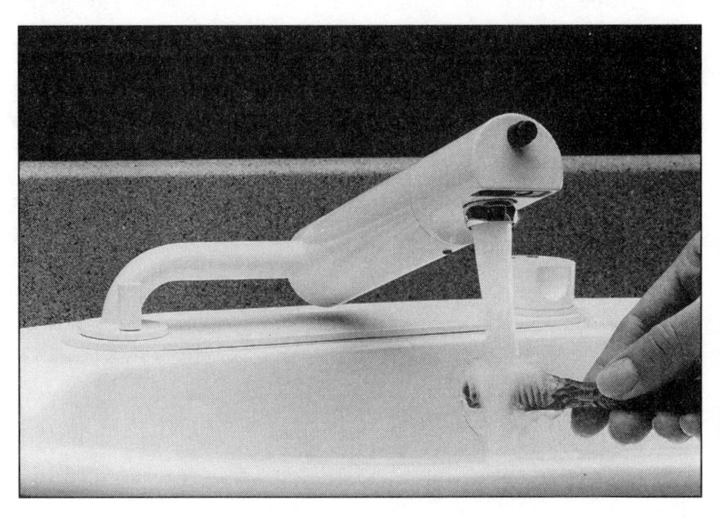

Infrared sensors control the flow of these Contempra Auto Water Faucets, delighting the Thomas Edison in us all.

The infrared controlled faucet is particularly effective for homes with young children who may tend to let water run, or for individuals who are physically disabled.

As with most new technologies, you have to pay to play. The Contempra infrared faucets start at $840. For further information contact Water Facets, 3001 Redhill, Building 5, Suite 108/145, Costa Mesa, CA 92626; 800-243-4420.

Point-Of-Use Heaters

The bathroom is probably the best place to consider installing a point-of-use hot water heater as a means to eliminate water waste. Several manufacturers have storage tanks and heating-element models available for easy installation under the sink. The water savings gained by these devices must be weighed with the cost of purchase, plus the cost of the energy needed for operation.

The Instant-Flow Tankless Water Heater, manufactured by Chronomite Laboratories, works with a faucet aerator to provide a half gallon of hot water per minute the instant you turn on the tap. Peak output temperature is rated at 40 degrees above incoming water temperature. Thus, 65-degree incoming water is delivered at a perfect 105 degrees.

The Instant-Flow heater must be hard-wired into your bathroom system but uses standard 110-volt household current. The cost for model S-30L for lavoratories is $250. For further information, contact Chromomite Laboratories, Inc., 21011 So. Figueroa St., Carson, CA 90745; 213-320-9452.

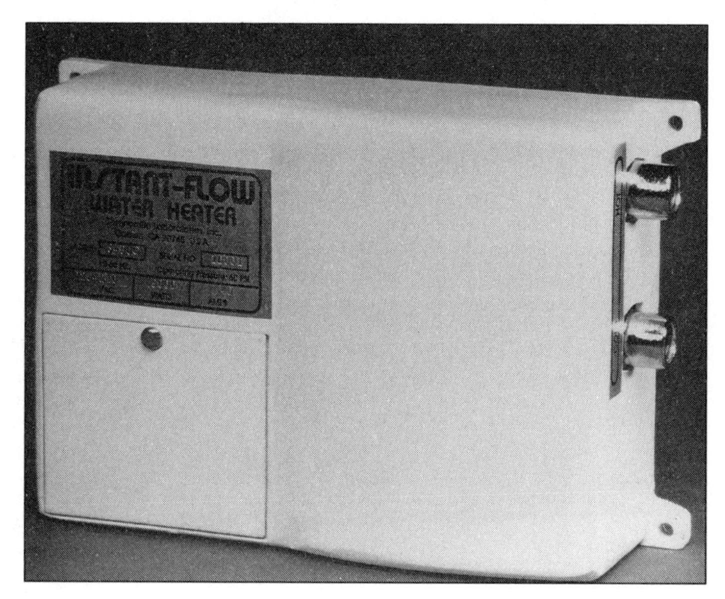

The Instant-Flow Tankless Water Heater provides a half gallon of hot water per minute while hot water arrives from the main water heater. No waiting, no water loss.

QuikFlo Point of Use Water Heaters keep 2.5 to four gallons of water heated under-sink, cost $188 to $198.

A second under-sink option, the QuikFlo Point of Use Water Heater, keeps 2.5 to four gallons of water heated and ready for use at all times. Thermostatically controlled, water temperature may be adjusted to a maximum of 140 degrees. One advantage of the QuikFlo system is that it can be plugged into a standard wall outlet, although special plumbing hookups are still required.

List prices are $188 for the 2.5-gallon model EQF250, $198 for the 4.0-gallon model EQF400. Contact Mor-Flo American, 18450 So. Miles Rd., Cleveland, OH 44128; 216-663-7300.

Kitchen

The average American spends quite a lot of time in the kitchen over the course of a year. And not just hanging around the hors d'oeuvre tray at parties, either. We're cooking, eating, washing dishes, eating, making drinks for the kids, eating, cleaning and hand-washing smaller items of clothing. All of this activity adds up to roughly four percent of the total household water used.

Saving water in the kitchen is not just a matter of buying the right appliances; it has as much to do with sensible, planned water usage by all members of the family. It's no good if mom and dad are careful about how long they leave the tap running and selecting the correct setting on the dishwasher if the kids let the sink overflow or run the dishwasher half full. It's important to make sure everybody using the kitchen knows what the goals are.

Thanks to the latest appliances, a major change in lifestyle isn't always necessary in order to reduce water consumption.

Perhaps the best news is that thanks to the latest appliances, a major change in lifestyle isn't always necessary in order to reduce water consumption. Still, developing thoughtful water management techniques will also make a big difference.

Europeans have been leading the way in water conservation for decades. The results are seen in the range and versatility of their designer appliances. But American manufacturers are no strangers to market-driven product evolution, and many of these companies are now entering the field with water-conserving appliances of their own.

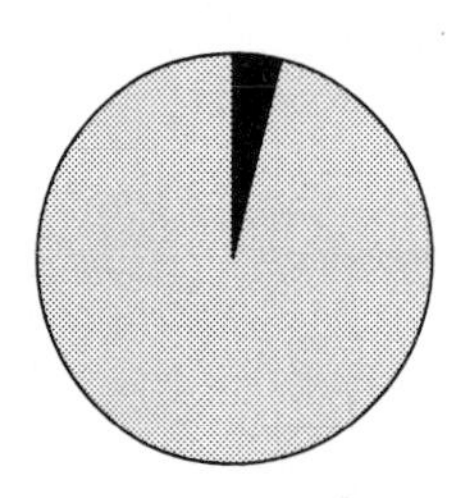

Kitchen 4% of home water use

If you're thinking about remodeling your kitchen, there's no reason you can't make economical water use as much a consideration as function, style and price when choosing appliances. And if you're not planning to remodel, there are still inexpensive accessories available which will make quite a difference.

One thing is for sure, there are plenty of ways and means to help us save water, energy and money in the kitchen. To help you engineer a water-smart kitchen, this chapter is organized into three main sections: Sinks & Faucets, Dishwashers, and Drinking & Cooking.

Sinks & Faucets

As you might imagine, most kitchen water usage takes place at the sink. Everything gets washed here: hands, food, tableware, and probably even the occasional baby or pet. Even if you have a dishwasher, you certainly hand wash dishes occasionally.

In addition, general household cleaning may start with filling up the sink before getting out the mops and cleansers. This places additional emphasis on the kitchen as a place water can be wasted – or saved.

This Sinks & Faucets section examines sink accessories and methods of using water that will be a big help in reducing water consumption.

According to Ohio State University, some 16 gallons of water are used to hand-wash eight place settings.

Hand Washing

According to an Ohio State University study, 16 gallons of water are used on the average to hand-wash eight complete place settings. Since the typical dishwasher uses only 12.3 gallons when set on its maximum-strength cycle, this leaves some doubt about the efficiency of hand washing.

There are various ways around the problem, but the best solution is found in a kitchen with two sinks. Use the drain stoppers and fill one with hot, soapy water for washing and the other with cold, clear water for rinsing. You should only need about five gallons total.

If you only have one sink, invest a few dollars in a plastic dishpan. Unsightly, yes, but worth its weight in gold and you can always hide it away when not in use. Besides, a dishpan can add a touch of authenticity to an otherwise unspoiled designer kitchen.

Water Smart Savings: 11 gallons per washing.

Come Together

A bowl or dishpan triumphs again when it comes to washing your fruit and vegetables. Instead of washing food as you need it for each meal, wash fruits and vegetables weekly before storing them in the refrigerator. Time consuming it may seem, but in fact you will realize time and water savings with this method. One quart will do for all the produce you want to wash. And think how nice it'll be to have clean veggies ready the next time you want to prepare a salad. Just chop and go.

Water Smart Savings: Four gallons per week.

Washing the week's fruits and vegetables at one time in a bowl in the sink saves time and water. The entire job takes just a quart of water.

Make A Daily Wash Tub

Think about how many times a day you run the kitchen faucet in order to just rinse off a simple utensil. As you're scrubbing the blade of a knife under the water stream, perhaps as much as a half gallon of water may be plunging down the drain.

Instead, why not each morning fill one side of the sink with soapy water and use this for the occasional wash job during the day? This will save water and the time required to accomplish myriad little chores.

Water Smart Savings: Five gallons per day.

The Price Of Leaks

A constantly dripping tap is enough to send anyone over the edge but it is surprisingly wasteful too. You can lose around 2,500 gallons per year just because of a worn washer. Most leaky taps are easy to fix. See the bathroom section beginning on page 13 for specific leak-fixing strategies. But if it's still all too much, call a plumber.

Water Smart Savings: Seven gallons per day.

The Essence Of Flow

Different areas of the household have different needs insofar as water flow is concerned. For example, the bathtub needs a faucet that will deliver as much water as possible in a short time, while a bathroom sink needs a faucet that will provide good washing effectiveness with the least amount of water.

Kitchen faucets fall somewhere in between, needing to speedily deliver water for cooking and yet not squander it during simple washing tasks. In many kitchens, turning on the faucet may well yield a flow of from three to four gallons per minute – hardly miserly stuff.

Before you buy anything, find out what's really going on at your kitchen faucet.

But before you buy anything, make sure you don't have a flow restrictor already. The easiest way is to look for a figure such as "2.0 GPM Max" stamped on the side of the faucet head, signifying that the restrictor will flow a maximum of two gallons per minute. Generally, any figure of 2.75 GPM or less will signify that you already have a flow-restricting device.

To pursue the issue further, wrap some adhesive tape around the old faucet head and unscrew it with a pair of pliers. A light touch works best so as to not deform the head. Once the head is off, examine it to see if it has a plate on the back with small holes drilled through it. This plate – along with a screen on the front side – is a good sign that you already have a common flow-restricting aerator faucet head.

In the end, the best way to tell what's really going on at your kitchen sink is to measure the output of the faucet. To do so, place a 10-quart pail under the spout and fully turn on the water...and a stopwatch. After 30 seconds, stop the faucet and timer, and divide the number of quarts in the pail by two. That's the flow in gallons per minute. As an example, a reading of 4.2 quarts in 30 seconds would equal a 2.1 GPM flow – pretty thrifty.

Let this number (and the quality of the water stream) tell you whether or not to go in search of a new flow restrictor. As a rule of thumb, you should consider installing a restrictor if your flow rate is currently over 2.75 GPM, or if the water flow pattern is uncommonly poor.

Choosing A Restrictor

The exact flow rate of your faucet has a lot to do with your home's inlet water pressure. Practically speaking, greater inlet pressure means a faster flow at the faucets. Since the inlet pressure is determined by your water company and very likely a household regulator, there may not be much you can easily do to improve it. However, there is no shortage of ways to control the flow rate at the faucet.

Screw-on flow restrictors can reduce a wild 4.0 GPM flow to a fraction of its former splendor and are available at most hardware stores for only a few dollars. Aerator restrictors mix air with water to give soft, bubbly flow, while laminar flow restrictors form water into a clear, gentle stream. Which you select should be determined by your kitchen needs and personal preferences.

Flow restrictors are commonly available in 0.5, 1.5, 2.0 and 2.5 GPM models, but for kitchen use a 2.0 to 2.75 GPM unit will work best. Costs range from $2 to $7 apiece.

Easy Off

One of the problems associated with adopting water-saving practices is that they can be quite labor intensive, with a lot of reaching to turn faucets off and on. That's a lot to ask of yourself year in and year out. So for a person that spends a lot of time at the kitchen sink, a quick-shutoff faucet head can make sense.

The unit pictured here flows 1.2 gallons per minute with standard household water pressure when in the "on" position, and 0.1 GPM trickle in the "off" position. Flipping the wire control lever is fast and easy, adding the all-important convenience that's necessary for a successful conservation program. Note that the shutoff is designed to temporarily restrict water flow, not to replace the main faucet handle or handles.

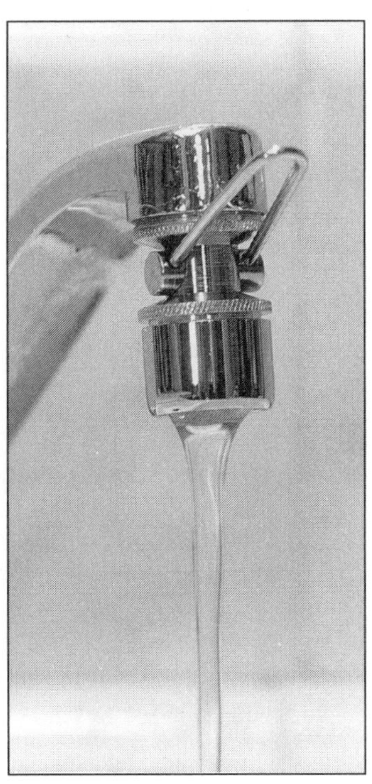

For about $10, this generic shut-off restrictor can reduce water flow 92 percent at the flick of a lever. Most positive aspect is ease of use.

Any number of methods may be used to estimate how much water such a device will save. But if it were to save 50 percent of water used during miscellaneous jobs at the sink, we're probably talking several gallons a day. The shut-off restrictor costs $10 at the local hardware store; similar versions should be available and competitively priced nationwide.

Water Smart Savings: About three gallons per day.

Pre-Set Water Flow

The heavy handed among us invariably push the lever type faucet on full force. Even with the traditional swivel-top tap, it's tempting to let it run full blast. Knowing it's wasteful is one thing, being more self-disciplined is another. And if you have kids in the house who love to play with water when you're not looking, this could be a losing battle.

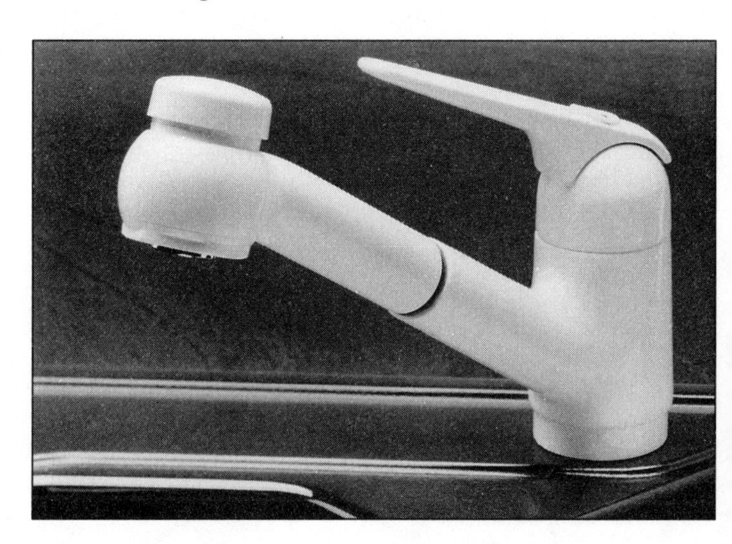

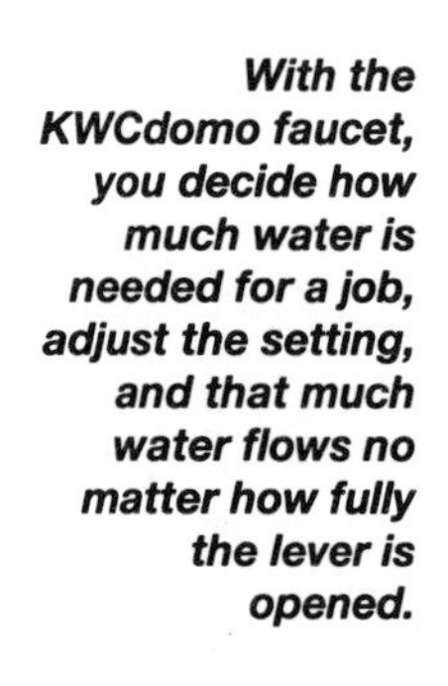

With the KWCdomo faucet, you decide how much water is needed for a job, adjust the setting, and that much water flows no matter how fully the lever is opened.

One answer might be a faucet which can be preset to deliver a chosen flow rate before you even turn it on. You simply decide how much flow you need for the job in hand, adjust the setting, and you've got the water flow desired no matter how far the lever is opened.

One such faucet is the KWCdomo. Swiss-made, it's expensive...up to $470...but has several other interesting features like seven temperature settings and built-in noise reducers. It features a retractable water wand with a spray head which can be adjusted from a strong spray to a strong stream. Find out more specifics from your local plumbing center or contact Western States Manufacturing Corp., 1559 Sunland Ln., Costa Mesa CA 92626; 714-557-1933.

Water Smart Savings: Up to 21 gallons per week.

Automatic Faucets

No matter how thrifty is a faucet's flow rate, leaving the tap running constantly is really the most wasteful practice. If you're a bit lazy about turning the faucet off while using it (that probably describes most of us), commercial-grade technology is coming to the home.

There are now available various faucets which use infrared sensors to detect any object that enters a beam's effective range. You may have encountered this system at public restrooms in airports or hotels. A microprocessor signals a solenoid valve to open and initiate the flow of water. Remove the object and the valve closes, causing water flow to cease. The future has arrived.

One manufacturer, Water Facets, has a range of such faucets called Contempra. They claim a whopping 80-percent water savings since water is flowing only when it's actually needed. Hygiene is improved too. Plus, you won't have to clean the faucet after washing your hands – covered in mud from the vegetable garden.

The high-tech Contempra faucet is turned on and off by infrared beams, reducing water consumption and leaving your hands free.

A continuous flow of water can be achieved by pressing a button at the head of the faucet. The solenoid valve is the only moving part; there are no washers or O-rings to complicate maintenance. The Contempra line is manufactured in the U.S. incorporating Swiss and Japanese electronics. Cost for the Jaguara kitchen system is $880. Contact Water Facets, 3001 Redhill, Building 5, Suite 108/145, Costa Mesa, CA 92626; 800-243-4420.

Water Smart Savings: Up to 80 percent.

The Kitchen Sink

If you can't stand the idea of the old-fashioned dishpan, today's sinks have an answer: a bowl and a half. Some chores might require deep water but not necessarily a lot of it. With a bowl-and-a-half sink, you can fill up the smaller bowl with less water than it takes to fill the bigger bowl halfway. This is a good alternative to the double sink unit. If you are ready to invest in a new sink, ask your local kitchen supplier for all the options available.

One premium example, the Kohler Bon Vivant model K5911, actually has three bowls, the smallest of which holds a thrifty gallon. Pricey at $432, it's nevertheless a perfectly simple way to wash or clean small items. For more information, contact Kohler Co., Kohler, WI 53044; 414-457-4441.

Water Smart Savings: Four gallons per week.

The Kohler Bon Vivant sink features a third bowl that holds a thrifty gallon, perfect for doing small jobs with a minimum of water.

Dishwashers

Even today some people consider dishwashers an extravagance. The machines use too much water, energy and people say they prove you're lazy. Owning a dishwasher doesn't prove you're lazy, but if your dishwasher doesn't have various cycle selections or if you use it carelessly, then the critics may have a point about the waste factor.

As the Ohio State University study points out, hand washing a load of eight place settings and serving pieces uses an average of 16 gallons. A standard dishwasher, on the other hand, will average between 9.2 and 12.4 gallons, depending on the cycle used.

If you are going to buy a new appliance, check out the water consumption rates and select a machine with the greatest choice of cycles.

If you are going to buy a new appliance, check out the water consumption rates for all cycles and select one with the greatest choice of cycles.

It's a fact that European and Scandanavian dishwasher manufacturers are generally more attuned to water conservation technology than their U.S. counterparts. On the other hand, American dishwasher companies are very competitive when it comes to offering electronic controls, different cycle options and "energy saver" features aimed at saving electrical power.

Fine-Tuning Dishwasher Use

Dishwashers represent a significant investment, but since 51 percent of American households own one, obviously we think they're worth it. To achieve maximum water efficiency with the cook's best friend, use the shortest running cycle you can to get each job done. The longer the cycle, the more water you'll use. Dishwasher owners, it seems, use the normal cycle 80 percent of the time while the short cycle is utilized for only 15 percent. Try the short cycle first for a normal load and see how clean it gets things. You may be pleasantly surprised.

Another tip is to avoid pre-rinsing. Hand rinsing isn't necessary for cleaning dishes in newer dishwashers. Scraping excess food off plates is enough. If the food is burned on, soak the dish for a while first.

Most important, wait until you have a full dishwasher before using it. Wasted space is wasted water. If you don't like to leave dirty dishes in the machine while you accumulate more, use the "rinse and hold" cycle, which consumes just a couple of gallons.

Water Smart Savings: Up to 50 percent.

You can maximize your dishwasher's efficiency by waiting until you have a full load before using it. Water savings can reach 50 percent.

The Mercedes Of Dishwashers

If you'd like a Mercedes-Benz but can't afford one, why not try the next best thing, a dishwasher built by the same company? AEG's Favorit 665i is claimed to use less water and less electricity than its rivals. The normal wash cycle, for instance, uses just 4.4 gallons.

One of the Favorit's unusual features is its use of 220-volt power instead of the standard 110-volt supply. The machine requires a cold-water hook-up only, and it begins its wash cycle with cold water to break down food proteins better than can hot water. Then the 220-volt system comes into its own by rapidly heating water to sterilize the dishes.

With less water use comes less energy use, says AEG. Plus, the Favorit is said to fully dissolve its detergent for greater environmental responsibility.

Attractive for its frugality with water, the AEG Favorit is nevertheless a high-line item at $1,614. For more information, contact Andi-Co Appliances, Inc., 65 Campus Plaza, Edison, NJ 08837; 800-344-0043.

Water Smart Savings: Seven gallons per load.

The Swedish Connection

Asko Inc. says that its Asea dishwashers have taken the lead in home water conservation. Made in Sweden, this product uses only 4.7 gallons of water in its normal wash cycle. Pots and pans require 6.3 gallons. This, reports Asko, saves 1,460 gallons per year compared to most top-of-the-line U.S. models, if you wash one load of dishes each day.

Triple filtration system of Swedish Asea 1502 dishwasher helps clean dishes with less water. The manufacturer's suggested retail price is $999.

Credited with helping to require less water is a triple filtration system that ensures food particles will not be redeposited on clean dishes.

The Asko Asea model 1502, which costs $999, has more than 20 wash programs, three temperature settings and room for 14 complete place settings. For information, contact Asko Inc., 903 Bowser, Suite 170, Richardson, TX 75081; 214-644-8595.

Water Smart Savings: Seven gallons per load.

Miele Magic

In 1929, Miele introduced Europe's first electrically-powered dishwasher in Germany. Today, its models are among the most water efficient. The company focuses on environmentally friendly appliances which need less water and detergent.

Miele's technical data shows that of five models, four use only 6.3 gallons, while one uses 6.7 all on the normal program.

Miele dishwashers save on detergent as well as water, cost a kingly $1,795 to $2,395.

For the pots and pans cycle, eight to nine U.S. gallons are used. There are seven to 10 wash programs, depending on the model.

They are able to reduce the amount of water used with the help of a powerful circulation pump and the triple filter system. The large, fine and micro-fine filters ensure that the recirculated water remains micro-filtered clean.

Manufacturer's suggested retail prices range from $1,795 to a princely $2,395; which, of course, will buy quite a lot of water. More information may be obtained from Miele Appliances, Inc., 22D Worlds Fair Dr., Somerset, NJ 08873; 800-843-7231.

Water Smart Savings: Six gallons per load.

American Classics

For the most part, our research has revealed that U.S. dishwasher manufacturers have to date concentrated more on reducing energy consumption rather than reducing water needed per cycle.

Whirlpool is a notable exception, offering the Power Clean feature that uses just 8.7 gallons for a normal cycle...about 2.4 gallons less than the average dishwasher. A low energy cycle uses only 6.5 gallons to clean lightly soiled loads.

The Whirlpool can save money in other ways: A delayed-wash function can retard the start of the cycle from one to nine hours, so you can load the machine and set it to start during off-peak energy cost hours. Prices range from $299 to $500. For more information, call Whirlpool at 800-253-1301.

Water Smart Savings: Four gallons per load.

Drinking & Cooking

We're not likely to conserve much water by not drinking any. The same goes for our family and our pets, so drink up and enjoy. But over the long run, there are some ways to save drinking water that you may not have considered and new devices are coming onto the market to help your efforts.

You may think that there's not much we can do conservation-wise in this area, but we can help to save water while we cook our food by looking at some of the technology designed for that very purpose and taking note of the following ideas.

Bottled Water

One option for your supply of drinking water is to buy bottled water, a growing trend nationwide. This concept seems unlikely to save either water or money until you consider two points: the cost of installing a dedicated water filtration system at home; and that water *purchased* for drinking is much less likely to be squandered than water running freely from the tap.

In short, the fact that you have paid for the water will make you less wasteful in using it. You might leave a drinking water tap running but you're unlikely to pour half a bottle of purified water down the drain.

Water Smart Savings: Up to two gallons per day.

Water purchased for drinking is much less likely to be squandered than water running freely from the tap.

Hot Water Dispensers

When you heat water in the kitchen, perhaps for a cup of coffee or mug of soup, chances are you'll heat more than you want in a kettle or pan. What you need is instant hot water, in just the right amount and no more.

This is where hot water dispenser systems come in handy. The dispensers use electricity to continually heat a small volume of water and deliver it through a special tap by the sink at temperatures up to 200 degrees.

Obviously, such instant hot water systems can not only make life easier in the kitchen, but will also save amounts of water over the long run. How much depends on how long you stand waiting for hot water from your tap each day, or how generously you fill your stove-top kettle for tea.

In addition, NuTone reports that Iowa State University found that the NuTone HoTTap system uses 33 percent less energy per day than a 40-watt lightbulb and 80 percent less energy than to heat water on a range top.

Hot water dispensers cost less than $200. Manufacturers include Nu-Tone, Madison and Red Bank Roads, Cincinnati, OH 45227; 800-543-8687.

Water Smart Savings: Up to three gallons per day.

Cold Water Dispensers

If it's chilled water you want, then cold water dispenser systems are also available in many refrigerators. Having chilled water on demand will save water by avoiding the need to run the tap while waiting for the water to cool down.

Various manufacturers also offer dual systems which can dispense either cold tap water or 190-degree hot water from the same spigot. One advantage of such a system is that you can add a point-of-use water filter or a water chiller to achieve complete filtered cold- or hot-water capability from one spout.

In-Sink-Erator offers its model HC at a cost of $349. In-Sink-Erator, 4700 21st St., Racine, WI 53406; 414-558-5712.

Water Smart Savings: Two gallons per day.

Hot and cold water dispensers can be useful water savers, at a nominal energy cost.

Filtered Water

There are a number of systems for turning the so-so quality water you get from the city or your backyard well into decent drinking (and cooking) water. It's important to be careful, though, in your choice of filtration system because some are surprisingly wasteful.

The reverse osmosis process, for instance, does an excellent job of removing impurities in water but in some cases uses up to three gallons for every one gallon of water produced. This is because most systems continue to process water whether it's used or not. One reverse osmosis system that avoids wasting so much water by incorporated a shut-off valve is made by Kinetico of Ohio. Although not quite as effective as reverse osmosis, carbon filter-based purifier systems are cheaper and produce drinking water without waste.

If you opt for a filtration system, remember to let the kids know that they can drink the water as soon as it starts to flow; since it hasn't been sitting in a pipe there's no need to let it run.

Chill Out

If the cost of a water dispenser system is outside your reach, then consider the simple expedient of cooling the water in your refrigerator. Instead of running the tap to get cold water for a drink, fill up a container and chill it in the refrigerator. Anything that prevents running that tap will save a couple of gallons daily.

Water left in the refrigerator in an uncapped vessel may pick up an unfavorable taste after a while. To combat this, store it in a capped container or else add fruit slices to the pitcher.

Water Smart Savings: Two gallons per day.

Instead of running the tap to get cold water for a drink, fill up a container and chill it in the refrigerator.

To Compost Or Not

According to the people at In-Sink-Erator, using your garbage disposer unit accounts for about 1.7 percent of your total household water consumption. So, when you've finished peeling vegetables and fruit for a meal, you might consider throwing the remains onto a garden compost heap. That is, if you have a big yard.

At an average water consumption of 1.1 gallons per use, the garbage disposer requires less water per day than one flush of a toilet.

Otherwise, the disposer may be a pretty good option after all. At an average water consumption of 1.1 gallons per use, the garbage disposer probably requires less water each day than a single flush of the standard family toilet.

The Last Straw

Instead of using a half gallon of water to boil a batch of vegetables into submission, try steaming them. You'll use far less water and the vegetables won't lose as many nutrients either. Either way, when you're finished you can still use the water for a great soup stock. Be sure to freeze it if you don't want to use it right away.

Water Smart Savings: Not much, but a moral victory.

Laundry

We are a tidy lot. At least, that's what Procter & Gamble says. The Cincinnati detergent company advises us that the average American consumer uses a washing machine 7.3 times each week, or some 380 times yearly. And with the average washing machine using up to 43 gallons of water per load, one family's annual clothes-laundering water use can easily surpass 10,000 gallons. Thus, laundry accounts for some 10 percent of our daily household consumption. The crummy part is that, unlike splashing around in a pool or taking long showers, doing laundry isn't even fun.

Although doing laundry ranks in the Top Three of water-using home activities, there are actually plenty of ways to do it with less water. As you've probably discovered elsewhere in this book, the methods involve both technologies and methodologies. You be the judge which works best.

Although doing laundry ranks in the Top Three of water-using home activities, there are plenty of ways to do it with less water.

Washing Machines

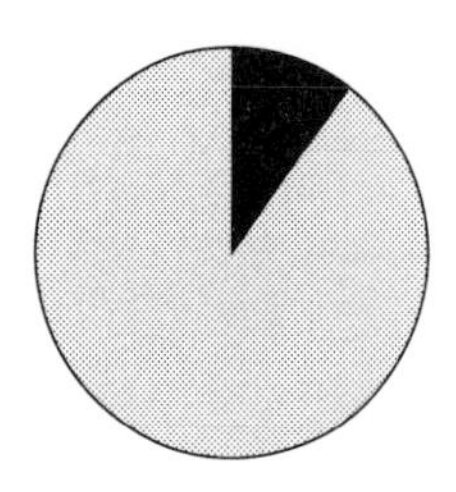

Laundry
10%
of home water use

Few of us are ever likely to go back to washing clothes on a rock. Dependent on appliances as we are, it therefore makes sense to consider the new wave of water-smart washing machines to help us get a handle on laundry water use. These machines use as little as 28 gallons instead of the traditional washer's 43 gallons per maximum load.

As usual, technology costs money...in some cases a *lot* of money. So before you leap, make sure you've maximized the efficiency of your existing washer. Then take a look at the cost of a new water-efficient machine spread out over a 20-year or longer life span. With reduced water and energy use, prolonged reliability, increased ease and enjoyment of ownership and better trade-in value, a new appliance might very well make sense.

Work To Capacity

Setting the load size when you do laundry can save 20 or more gallons of water; deleting the second rinse another 24 gallons.

If you already own a washing machine, chances are good that it has an adjustment for load size. This simply controls the amount of water that enters the machine. A smaller load of laundry needs less water, right? So remember to set the load size when you do laundry. Doing so can save 20 or more gallons of water when you have only a few clothes to wash.

If your machine doesn't have a load size setting, it will use the maximum amount of water every time you wash...even if the load is only a pair of socks. So make sure you always fill the machine to capacity with laundry.

It turns out that doing full loads of laundry is the preferred conservation method anyway. The appliance makers we've spoke with have told us that machines actually use water most efficiently when they're working at maximum capacity.

Water Smart Savings: 20 gallons per load.

You Only Rinse Once

Some washers also offer an optional second rinse cycle in addition to different load-level settings. The purpose of the cycle is to make sure dirty laundry comes out squeaky clean. Anyway, you may be able to tell your machine to forget the second rinse, thus saving some 24 gallons of water. You shouldn't need the second rinse anyway, unless you've just replaced the clutch in your pickup and your clothes are miserably dirty.

Look on your washer's control panel. If it doesn't have a rinse-delete feature, you can cancel the second rinse by manually advancing the dial to "Off" once the primary wash/rinse cycle has

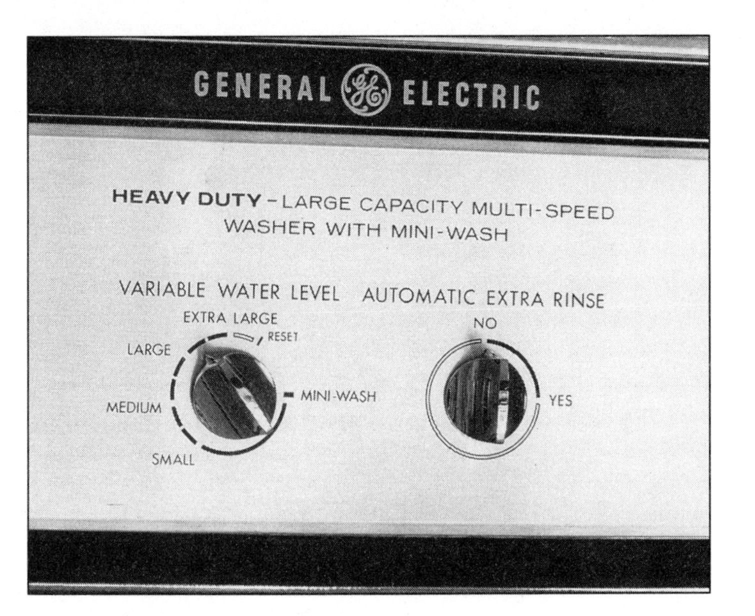

Job One for efficient laundering is to set your machine properly. Simply stated: wash large loads only, without the extra rinse cycle.

been completed. Keep a portable timer handy to remind you, or develop a telepathic relationship with your washer.

Water Smart Savings: 24 gallons per load.

Water-Smart Washers

As usual, throwing money at a problem can make a big impact, so here goes. A new wave of front-loading Euro-style washing machines is now available from domestic as well as European manufacturers. These machines are designed to use about 30 percent less water than normal washers.

An outlay of $649 to $1,738 will get you one of these new machines and the envy of your neighbors. Of course, any washer costing this much will take a long time to pay for itself in water or electric savings, but eventually it will. Just how long depends on how often you do laundry, but if a big family indeed does an average of seven loads every week, one of these new washers will save on the order of 110 gallons per week, 475 gallons per month and 5,700 gallons per year.

Water costs vary from region to region, but at a nominal cost of $1.85 for 748 gallons (one "unit"), you're still looking at something like 46 years to pay off even the cheapest front-loading machine with water savings alone. Obviously, you'll have to count energy savings as part of the monetary equation...even though nobody we know wants to do laundry for anywhere near a half century.

The following machines appear to be sound possibilities, but by all means consult with *Consumer Reports* and other authorities, compare features, and form your own judgements before purchasing. Appliance manufacturers are sometimes reluctant to set retail prices nationally, so be sure to shop dealer-to-dealer for the best combination of service and price.

The AEG Way

AEG bills itself as the German company that builds "environmentally friendly appliances." Its front-load washers use a total of 29 gallons on even the longest cycle, for a savings of 14 or more gallons per load. The Lavamat 539 and Bella Super models also heat their own water via a 220-volt electrical hook up, which is required.

The advantage of the self-heating process, says AEG, is that the wash starts with cold water to break down proteins during the filling process. Then the washer heats the water to sterilize the clothes. This method not only eliminates the need for bleach in most cases; it also will not deplete the hot water heater no matter how aggressively you launder.

Finally, an exclusive "ECO lock" feature is designed to seal off the suds container during the wash cycle, preventing any loss of detergent. As a result, AEG owners are advised that they can use 20 percent less detergent than recommended by soap companies. Additional information on AEG's $1,500 Lavamat 539 and $1,600 Bella Super washers may be obtained from Andi-Co Appliances, Inc., 65 Campus Plaza, Edison, NJ 08837; 800-344-0043.

Water Smart Savings: 11 gallons per load.

The GE Mini-Basket rides on the main tub shaft, washing small loads in just 11 to 14 gallons of water.

Do The Mini-Basket Thing

Suppose you want a super-capacity top-loading washer and will need to do small loads of laundry anyway? General Electric offers a Mini-Basket tub feature on most of its washers designed to take care of such dilemmas easily and efficiently.

The GE Mini-Basket setup consists of a special wash basket that rides on the main tub shaft, providing a separate environment for small loads or delicates. Used in conjunction with the Mini-Quick Cycle found on GE's top washers, the Mini-Basket can do small load of laundry in just 14 minutes with 11 to 14 gallons of water.

GE models WWA9890M, 9850M, 9800M and 8800M offer both the Mini-Basket and Mini-Wash features, and cost between $499 and $699. Contact your local dealer or call GE at 800-626-2000.

Water Smart Savings: 26 to 29 gallons per load.

12 Percent Smarter

Kelvinator top-loading washers are described by the company as sensible, dependable and affordable. How well these claims hold up may be at least partly determined by examining the company's product specifications chart. Models AW300G and AW700G are said to use 22 to 43 gallons per wash load compared to 49 gallons for previous models. Kelvinator credits the 12-percent savings to a Double Scrub System that makes wash action more efficient.

Singles or small families may find that a portable washer provides attractive water and monetary savings.

A more aggresive water-conserving machine is the Kelvinator portable model AW330G, which uses only 17 to 31 gallons per load. Singles or small families may well find that such a washer provides attractive water and monetary savings.

Prices are approximately $448 to $465 for models AW300G and AW700G; and $498 for portable model AWP330G. If you can't find a local Kelvinator dealer, call the company at 800-323-2440.

Water Smart Savings: Six to nine gallons per load.

Premium Choice

Another German manufacturer with unique thoughts about laundering is Miele. Its model W 1070 front-loading washer uses between 21 and 32 gallons of water per load, depending on which of 14 available function programs is selected.

In addition to saving water, the Miele is said to remove dirt more efficiently than traditional washers due to its cold-water fill process, which soaks out organic dirts before gradually heating water up to remove inorganic dirts. An extensive control panel also allows you to precisely set water temperature between 80 and 200 degrees, providing energy savings.

Shown here with matching dryer, the Miele W 1070 front-loading washer uses as little as 21 gallons per load, costs a lot: $1,738.

This is a premium washer, no doubt...the Miele W 1070 costs $1,738. More information may be obtained from Miele Appliances, Inc., 22 D World's Fair Dr., Somerset, NJ 08873; 201-560-0891.

Water Smart Savings: Eight to 19 gallons per load.

Sears' Solution

Compared to Sears' own top-loading machines, the Kenmore model 49881 front-loading washer can save some 15 gallons of water per load. This 35-percent reduction (the washer uses 28 gallons for a full load) comes from a front-loader's inherent

Sears Kenmore model 49881 washer saves water with a low 28-gallon consumption for large loads. Euro-styling and low $649 price make an appealing choice.

water efficiency, not from downsized performance. Sears in fact claims that its front-loading machine cleans clothes effectively thanks to a basket that alternately rotates clockwise and counter-clockwise. The price is $649.

Water Smart Savings: 15 gallons per load.

The Suds-Saver

Maytag offers several different Suds-Saver machines that can save their soapy wash water for use in successive loads of laundry. This saves not only water, but detergent. It's a great idea that has been around for years.

When the Suds-Saver feature is used, the Maytag pumps its 16 to 19 gallons of soapy wash water into an auxiliary holding tank after the wash cycle. The machine then progresses to rinse and spin cycles. As the next load of laundry begins, the stored soapy water is pumped back into the machine for a second use. "Save" and "drain" buttons allow you to control when you wish to use the Suds-Saver feature.

The exact amount you'll save depends primarily on the size of your laundry loads and how many times you wish to reuse the soapy water. Typically, one load of sudsy water can be used a

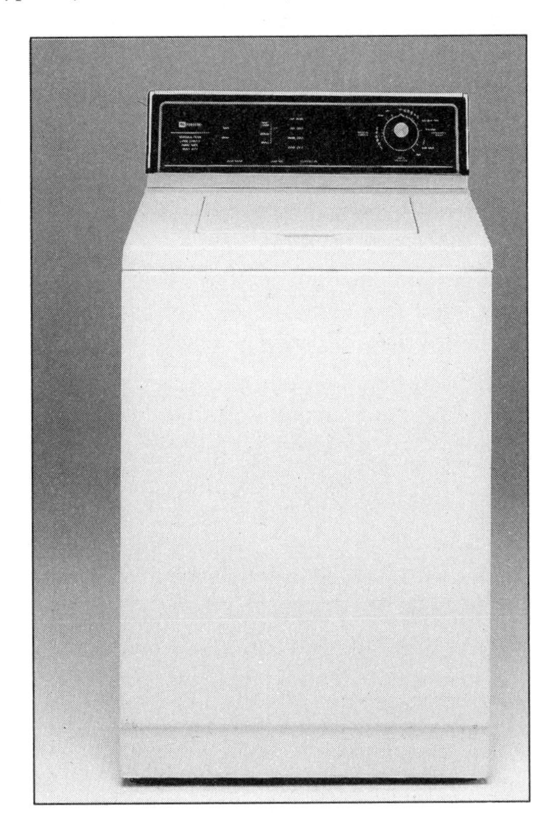

The Maytag Suds-Saver washer stores and reuses 16 to 19 gallons of sudsy water for each load of laundry.

maximum of three times. To use an example provided by Maytag, the Suds-Saver feature can save 32 gallons of water and two cups of detergent for every three large loads.

The Suds-Saver feature is available on top-loading Maytag models A7500, A9400 and A9700 large and extra-large washers, which cost between $549 and $628. To learn more, contact a retail outlet or Maytag Company, One Dependability Sq., Newton, IA 50208; 515-792-8000.

Water Smart Savings: 16 to 19 gallons per load.

Grey Is For Great

Even if you don't want to invest in a high-dollar European washer or a domestic suds-saver machine, you can still harness the wash or rinse water coming out of your washer for further use. Since the average machine uses 43 gallons of water per full load, that's a substantial amount over the course of a year...no matter how much laundry you do each week.

The way to do it is to build your own laundry room grey water collection and pumping facility. Here's how. Go to your hardware store and buy a 44-gallon plastic trash can (stronger is better), a submergible electric pump (110 volt), a float switch to shut the pump system off when the trash can is empty, and 50 feet or so of garden hose. Total cost with all-new components should be well under $200.

Assemble the components with the hose attached to the submersible pump, the pump and float switch inside the trash can, and the outlet from the washing machine aimed into the trash can. Now when your clothes are finished washing, the wash and rinse water (with biodegradable soap) will be dumped into your new holding tank. At this point you can use it for all your grey water gardening needs – or as a make-do suds-saver for additional laundry loads – merely by plugging in the pump.

With the average machine using 43 gallons for a full load, there's a substantial amount of grey water available over the course of a year.

The system shown on the next page can empty its 44-gallon trash can in seven minutes at a flow rate of about six gallons per minute. That's the same as a typical garden faucet.

Here's the list of components that went into our working model:

- Heavy-duty 44-gallon seamless trash can such as the Rubbermaid Brute, $39 at hardware stores.
- Little Giant Water Wizard pump model 5-MSP, about $90 or less at hardware stores. Little Giant Pump Co., 3810 N. Tulsa, Oklahoma City, OK 73112; 800-468-7867.
- Float switch model NM15K, about $28 at hardware stores. Simer Pump Co., P.O. Box 2973, Mission, KS 66202; 800-468-7867.

• 50-foot, 5/8-inch diameter "soft and supple" garden hose, about $27 at hardware stores.

One cautionary note before proceeding – first check laws regarding the application of greywater in your area. Greywater is not advised for irrigating food crops, and in some cases it may be used only in underground irrigation systems.

If you're more interested in a ready-made setup, the Simer Pump Company listed here also sells a complete laundry pump system that's ready to plug in and work. Model 2925 includes a submersible pump, a 10-gallon storage tank, and a float switch to control operation. Please note that this system starts automatically as soon as laundry water enters the holding tank. This means coordinating your garden's need for grey water with your need to do laundry. Cost for the complete Simer system is $199.

Water Smart Savings: 43 gallons per load.

For less than $200, you can put all your laundry wash and rinse water to use with a home-built grey water pump system.

Find A Grey Soap

Ecover and Oasis brand biodegradable soaps clean satisfactorily and may be safely used in the garden. A one-gallon jug of the Oasis costs under $25, and that's enough to clean between 80 and 128 loads of laundry at a cost of 19 to 31 cents each.

Biodegradable soaps make good overall sense for the environment, not just for helping to save water in conjunction with grey water systems.

If you can't find these soaps locally, contact:

• Ecover, Inc., 6-8 Knight St., Norwalk, CT 06851; 203-853-4166.

• Oasis, Biocompatible Products, 1020 Veronica Springs Rd., Santa Barbara, CA 93105; 805-682-3449.

Hand Washing

The world's first "washing machine" probably consisted of a tub filled with soapy water and clothes, which were then hand-agitated by the lucky resident homemaker. Well, the concept still has merit when it comes to home water conservation.

With a little imagination, water can be saved in the laundry area even without spending money or even much time or effort. You'll no doubt find the first idea to be the easiest, even if it's not the most aesthetic.

Washing clothes half as often can save the average family up to 157 gallons of water per week.

To Smell Or Not

Just let you mind soar for a minute. Imagine the time and water you'll save if you only wash clothes when they are truly dirty, say once instead of twice a week.

By cutting your clothes washing in half, you'll save 50 percent of the water, 50 percent of the laundry detergent, and 50 percent of the energy required to heat the water and dry the clothes. You'll also save 50 percent of the time you spend doing laundry in the first place.

Calculating how much water this will save means knowing how many loads we do every week. According to Procter & Gamble's 1988-1989 survey, the average household does 7.3 loads of laundry per week. At an average maximum water use of 43 gallons per load, this means that washing half as often can save up to 157 gallons per week.

The goal is not to make you look and smell like an old cow hand. Just realize that not every garment has to be washed each time you wear it. How to tell when to wash? There's no official formula...just let the old sniffer do a test. If you can smell it, wash it.

Water Smart Savings: Up to 157 gallons per week.

Tandem Tupperware

To do small amount of washing, set a couple of big Tupperware or similar bowls in the kitchen sink; one side for soapy water, one side for rinse. Not only can you now control the amount of water that's used, but you can use the leftover water on outdoor plants when you're done. It's fast, it's efficient (one gallon each for a small wash and rinse) and if biodegradable soap is used, your outdoor plants will love you for it.

The option to hand washing is to save up enough clothes for a full machine load, or to run your washing machine at less than peak efficiency. There goes up to 43 gallons down the drain. Although the Tupperware/sink method can save a lot of water, it's obviously no good if you've got cooking going on at the same time. And don't forget to wash and rinse light and dark clothes separately.

Water Smart Savings: Up to 41 gallons per load.

Grandma's Tub

What does a rural grandmother have in common with someone living on the 33rd floor of a Manhattan skyscraper? They both do laundry now and then with a washboard and tub. For grandma on the farm, these antiques may be a necessity; but others simply recognize the savings and convenience a washboard and tub present for small laundry needs.

If you're serious about saving water you can do it too, and it's not all that hard or time consuming. With a fraction of the water used by most machines, you can be done with a few clothes before your machine has even shifted into high gear. Using just four gallons to wash and rinse, a washboard and tub (or sink) thus make the perfect alternative laundry facility. When you're done washing with biodegradable soap, use the water in your non-edible garden.

Grandma's laundry formula still makes sense: a galvanized tub and Columbus washboard can do small laundry loads with just four gallons of water.

Was grandma water smart, or just stubborn? A galvanized tub and washboard are perhaps the ultimate water-saving tools; can do small loads with only four gallons.

The Columbus Washboard Company has been around for over 80 years, and the family business still sells 10,000 to 20,000 washboards per year to serious launderers in the U.S. and abroad. Another 120,000 or so are destined to be used for decorative purposes...which tells you what people really think washboards are for. A number of models are available, but model 2062 (brass and wood) and model 2080 (glass and wood) are recommended for durability when used with with harsh water or common household detergents. Costs range from $14 to $17.

A common 15-gallon galvanized steel tub should serve occasional laundering needs well and costs less than $20. You may be able to find both tub and washboard at Ace, True Value or other hardware stores. If you can't find washboards there, contact Columbus Washboard Co. at 614-299-1465 for the name of your nearest dealer. While you're at it, pick up a shawl and bonnet.

Water Smart Savings: Up to 36 gallons per load.

Yard, Garden & Pool

The lawn outside your home truly is a "field of dreams." Lawns and gardens present our biggest opportunity to save water and money, and the bigger the field, the bigger the dream.

The California Department of Water Resources estimates that the typical home uses about 150 to 200 gallons of water daily, which averages out to 64,000 gallons annually. Of this, nearly half is used outdoors. And as much as 40 percent of this can be wasted due to improper watering techniques – 11,000 gallons each year.

Watering only as needed can save as much as 1,200 to 1,500 gallons per month.

In addition, the Metropolitan Water District of Southern California reports that watering only when the lawn or plants show signs of needing it, rather than on a regular schedule, can save as much as 1,200 to 1,500 gallons per month.

How can this be? Consider that in a single hour, a lone sprinkler dispersing just five gallons each minute (only half the maximum flow rate of a 1/2-inch hose) will use twice as much water as ten toilet flushings and ten loads of laundry.

That shouldn't be too surprising, considering that the same 1/2-inch hose can pour out 600 gallons in an hour, and that a 5/8-inch hose can flow 1,000 gallons per hour. If you leave your sprinklers on all day, thousands of gallons can be wasted.

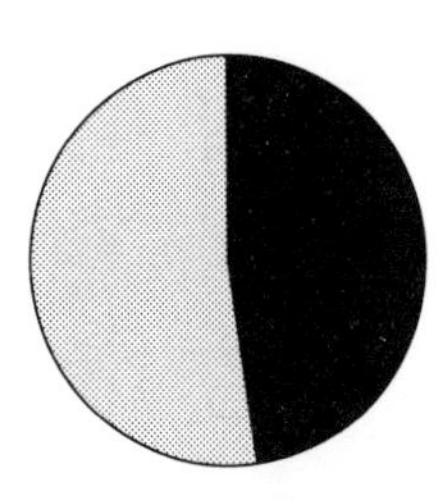

Yard & Garden 48% of home water use

But this doesn't mean you have to stop using water outside, have a brown lawn, and fall into poor repute with the neighbors. By applying a little common sense and adjusting your habits, you can eliminate water waste and still enjoy a healthy yard.

This chapter will show you how you can change the way you approach watering to reduce your water consumption by about one third. Methods include replacing thirsty bluegrass lawns with drought-loving plants and ground cover, installing an efficient drip irrigation system to encourage a healthier, less water-demanding yard, and reusing water and captured rain to save thousands of gallons annually.

If you have a swimming pool or outdoor spa, you can reduce heating and water bills by up to 70 percent for the price of a cover.

So don't view water conservation in the yard as a burden. Instead, celebrate an opportunity to put the money that you needlessly spend on outdoor watering into your bank account.

Celebrate an opportunity to put the money you spend on outdoor watering into your bank account.

Lawns

Compared to most plants, grass is a thirsty waterholic, requiring many times more water than shrubs and trees. No other single area around the household soaks up more water than lawns.

In water-stressed areas of the country, it's getting to be a common sight indeed to find lawns shrinking and native plant beds growing. Having less grass of course means less watering and mowing. We'll tell you how to achieve this while maintaining a lush and impressive landscape in the process.

If you crave acres of lawn, we'll point out methods and tools you can use to dramatically reduce the amount of water you pour onto the lawn.

We'll also give you dozens of tips on how save water all over the garden.

Know Your Flow

Before you set out to banish lawn sprinklers from your life, take stock of how much water they're using. Here's a simple way you can put the question of irrigating your lawn into perspective. First, turn on your sprinklers (be sure no other household water is running), and observe your water meter. Record the volume of flow after ten minutes.

Next, prepare yourself to receive some distressing information. Now multiply the number of gallons used in ten minutes by six to determine how many gallons of water you dump on your lawn in an hour. You may now enter a state of shock.

The realization that hundreds of gallons of water per hour can be distributed to your lawn certainly is significant compared to the indoor areas of your home – even the bathroom. Of all the kingly ways to use water, lawns are the most notorious.

The household water meter is a faithful servant that may be relied upon for accurate information about how much water your lawn uses.

Applying Water Smartly

"Irrigation" is really just a fancy word for applying water, and for people who hate to spend money on watering the ground, to irrigate is to irritate. Therefore, the primary water-saving approach to irrigating your lawn is simply knowing when it actually needs water, and applying the exact amount consistently. Thus, the first law of lawn water conservation: water only when you must.

Grass, like most outdoor vegetation, is fairly hardy stuff and can easily survive extended dry periods. In drought conditions it can go into a semi-dormant state and be revived later. So give your turf credit for being durable in the presence of dry conditions.

Allowing your lawn's soil to dry out, and then watering it generously with a slow trickle to get to the roots, is good for it. On the other hand, frequent water applications with light sprinkling can actually harm it.

Train Your Roots

If water sinks in only a few inches, you'll end up with shallow roots that lie close to the surface. This actually weakens the plant, making it more susceptible to diseases and more fragile in dry conditions and during winter. It also causes a loss of nitrates in the soil, an invasion of weeds and soil compacting.

To help eliminate these dastardly conditions, you can "train" your root system. By soaking the lawn enough to get water deep into the ground, you get to the root of the problem, so to speak. This encourages the roots to grow deeper – closer to moisture deep in the soil and away from surface evaporation. It also makes the plant stronger and more resistant to a variety of nasty conditions such as drought, disease and winter.

Delaying watering during the first weeks of spring encourages deeper roots, delays the need to mow, and makes a healthier lawn. For the sake of your lawn (and water bill), don't be a watering fiend. In sum, don't worry about killing your grass if you don't water it. Let the soil dry out a little, then soak it deep.

Water Smart Savings: Up to 50 percent.

Dig For The Dirt

Plants have been growing on this earth since dirt was news. There are many types of dirt, ranging from dense clay to loose sand. In between is an ideal loamy soil that plants thrive in.

Knowing the makeup of soil in your yard is central to understanding when and how much to water.

The type of soil that your plants grow in determines how fast water is absorbed and stored. In short, outside of other environmental factors – like whether the mercury is hovering near 105 – dirt type is the key element that determines how and when you must water.

Clay absorbs water slowly, sometimes at a rate of only 1/4-inch per hour, before water begins to run off. The flip side is that clay can hold large amounts of water for long periods. Since water runs off easily, water must be applied as slowly as possible, sometimes in spaced intervals. For example, you might water clay for ten minutes, then wait 20 minutes for that to absorb before adding more. You might need to water clay once every week or two.

Contrarily, sandy soil absorbs water quickly, sometimes at a rate of two inches per hour. Since water runs easily through it, you can apply water quickly. However, because sand can't retain much water, you'll have to water more frequently. If your soil is very sandy, you may have to water every few days.

Loamy soil is ideal. It contains the optimum amount of sand and clay, and has the ability to absorb and store water well. Plants love loam.

Assessing Soil Type

Knowing the makeup of soil in your yard is central to understanding when and how much to water. To learn your soil's content, you'll have to take samples. But before testing your soil, water it enough to be able to make a moist ball of soil, not too dry or muddy.

Squeeze the soil between your fingers to ball it up. Clay feels greasy and slick, and forms ribbons when rolled. Sandy soil feels loose and gritty, and cannot be formed into a ball, no matter how artfully you manipulate it. Best, loamy soil feels gritty, moist and substantial, and can be packed into a ball that falls apart easily.

For a complete soil analysis, you'll have to take a deeper sample to a lab. And to get at the deep soil you can either dig a big hole in your yard and alarm the neighbors by re-enacting a scene from *The Great Escape,* or use a more civilized method involving a soil sampler tool. This is simply a device you push in the ground to extract a core of soil.

One such soil sampler is available from Unique Landscape Necessities. Model 0888 costs under $25 and may be ordered direct from UNL. To obtain more information or to order, contact Unique Landscape Necessities, 5733 Ocean View Blvd., La Canada, CA 91011; 818-957-0188.

Moisture Content

Regardless of what type of soil you're dealing with, knowing how much moisture is contained within it will pay dividends in plant health...and lower water bills. To find out, you can use a truly professional soil probe that comes complete with a carrying case. Now you can go out in the yard, adjust your spectacles and clipboard, and probe the soil for immediate and accurate moisture content. It's a must for large properties that have equally large water bills.

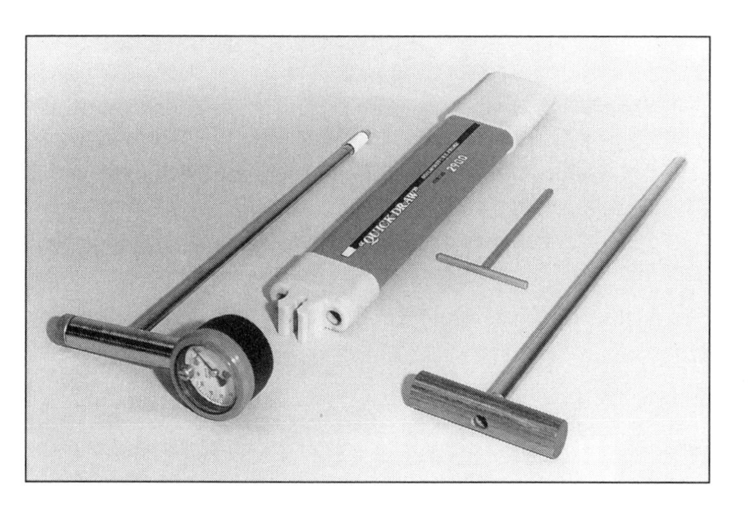

The Quick Draw soil moisture probe costs $235, determines the underground moisture content anywhere on your property within minutes.

Soilmoisture Equipment offers its Quick Draw soil moisture probe model 2900FI to suit. Priced at a healthy $235, the probe is designed to quickly let you know how much water any soil is carrying, anywhere on your property. After making a hole for the probe with the supplied coring tool, the probe is inserted into the ground and allowed to measure the "soil suction," the force a plant must exert to soak up water.

Besides helping you determine when to water, a tool of this sort can help you understand why certain areas of your yard provide superior or inferior plant growth. For more information on this versatile tool, contact Soilmoisture Equipment Corp., 801 So. Kellogg Ave., Goleta, CA 93117; 805-964-3525.

Water Smart Savings: Up to 20,000 gallons per year.

Knowing When To Say When

You already know how to determine how much water your sprinkler system can pump out in an hour. Here is a simple way to determine the bare minimum amount of water your lawn needs, and to assure that this is all it gets. Begin this study on what you perceive to be an average lawn-watering day: i.e., not during a monsoon, and not after last night's watering extravaganza.

Equipped with a soil probe, watch, rain guage and clipboard, you can act as your own lawn irrigator.

First, write down the reading of your water meter. Second, turn on your sprinklers and keep track of the watering time with a stopwatch. Third, place a few inexpensive rain gauges around the lawn and note on the average how much water collects.

During this watering cycle, you can also periodically take soil samples with a soil probe to determine how deeply the water has penetrated the root system. When the water has reached a sufficient depth to soak the roots, note your meter reading and the time. You now know how much water – and how much time – is required for optimum watering of your lawn.

With this groundwork under your belt, you have several new methods of accurately controlling the amount of water used on your lawn:

1. Run the sprinklers for the same amount of time required to soak the roots in this experiment.

2. Run the sprinklers until the water meter reading duplicates the amount used in the test.

3. Run the sprinklers until the rain gauges fill to the ideal levels established in the test.

4. Use a soil moisture probe to evaluate moisture penetration each time you water. Note: you'll discover that soil conditions vary greatly as a result of humidity, temperature, wind, and sunny or cloudy skies.

5. Use two or more methods simultaneously.

Congratulations. Equipped with a soil probe, watch, rain gauge and/or clipboard, you now qualify as a trained lawn irrigator. You'll always have a way to apply the exact amount of water that's necessary to nourish a grass roots movement. Celebrate with a root beer.

Water Smart Savings: 18,000 gallons per year for a 30 x 40-foot grass plot.

Step On It

The question is, how can we know when our lawn really needs water without sinking the price of a Las Vegas weekend into special tools? A simple and free indicator is to take a walk on it. If the grass springs back up reasonably quickly, say within a minute or so, don't touch the sprinkler knob. If the grass lays flat and you can see your footsteps even after a few minutes, it's probably time to give it a drink.

Another sign of a thirsty lawn is a dull green color. In some areas, like along sidewalks, this color might appear more commonly than it does in other spots. Remember, it's only necessary to water areas that show signs of thirst, not the entire lawn.

Water Smart Savings: 40 to 50 percent.

Mowing Tips

A summer tip. In hot weather, let the grass grow a little taller, because tall grass retains moisture better than short grass. It's more fun to walk on too. Also, leaving rather than collecting the cuttings adds nutrients to the soil. The nutritious clippings will help to keep the underlying dirt moist and cool.

Scheduling Irrigation

The best time to water your lawn is early in the morning. By watering early, the soil will absorb more of the water you apply because so much less is lost to evaporation.

It's only necessary to water areas that show signs of thirst, not the entire lawn.

It turns out that watering during the day can actually harm your lawn. Water drops clinging to the blades do a fine job of acting like little magnifiers, causing scalding and burning to the grass blades.

Watering in the evening is the second best time, but keep in mind that evening's cool, wet conditions may well be just right for a lawn fungus to start.

Lastly, it's a good idea to postpone watering when the wind is blowing. Wind carries water away and accelerates evaporation.

Water Smart Savings: Up to 40 percent depending on temperature and humidity.

Rain Checks

The formula you've worked out for watering your lawn will need to be modified periodically depending on season as well as prevailing conditions. When rain falls, the same inexpensive little rain gauges you used to check your sprinkler coverage will help you quickly know how much water to subtract from your lawn's weekly ration. They only cost about $4, and may even be available for free at your local water department.

Inexpensive rain gauges are available from a variety of sources, tell how much water to subtract from our watering plan after a rainstorm.

Reducing Sprinkler Waste

Consider for a moment the way in which your sprinklers actually distribute water. Oscillating sprinklers that throw water in high, fine sprays are inefficient because much of the water can be lost to evaporation and to the wind. A sprinkler that throws large droplets in a flatter pattern delivers more water with less waste. Thus, adjusting or replacing the sprinkler heads can reduce evaporation and wind losses.

Water Smart Savings: Up to 20 percent.

The Seven Commandments Of Lawn Watering

Use the following list to help you minimize your water use while maximizing the health and beauty of your lawn.

1. Water only when you must.

2. Deep-soak the roots; daily sprinkling can damage grass.

3. Water in the cool of early morning.

4. Avoid watering on windy days.

5. Aim your sprinklers, and don't waste water on open dirt, sidewalks or driveways.

6. Adjust your sprinklers to throw a low pattern of water to minimize evaporation.

7. Monitor the amount of water you apply.

Lawn Alternatives

You no doubt have a legitimate reason for wanting to maintain a lawn. Most probably, that reason is that the lawn was there when you moved into your home. Plus, lawns are fun to play or lounge around on, while succulents like cacti are not.

Replacing lawns with attractive hedges or flower beds creates a lush and inviting alternative to watering – and mowing.

But do you really need a large lawn? If not, you have many attractive alternatives. By reducing lawn acreage and increasing foliage, you can create a pleasant walkway that meanders through a garden. You can also make a more interesting landscape, along with a better one for wildlife. So consider replacing part or all of your lawn with hardy flowers, hedges or ground cover.

Water Smart Savings: 2,500 gallons per year for a 3 x 30-foot section.

The Case For Patios

A gazebo, patio or deck doesn't need watering or mowing. Although these things may cost plenty to install, they can also add value to your property. Amortized over the years, they will definitely be less work and less water-intensive than grass. Finally, these options are worth considering because they enhance your yard by breaking up the monotony of flat fields of green.

Water Smart Savings: Up to 5,400 gallons per year for a 10 x 20-foot deck or patio.

Zeroing In On Xeriscape

Another alternative to water-hungry lawns is xeriscape. Literally a combination of the Greek word "xero" for dry, and the English word "landscape," the xeriscape is an unthirsty landscape that requires little maintenance. The conversion process, insofar as lawns are concerned, involves installing water-sipping plants in the place of water-guzzling grass.

Xeriscapes incorporate plants indigenous to your area to replace water-hungry specimens.

Part of being water smart is recognizing which plants grow wild in your area. Some of these are bushy growths commonly known as weeds, while others are flowering plants called wildflowers. Regardless of whether they're called weeds or wildflowers, don't let the labels fool you: the right native species can be both attractive and economical. So why not consider letting wild growth prosper, rather than weeding it out?

Other flowers and shrubs that are indigenous to your area will require only the average regional precipitation to survive. That's the wonder of natural selection. There are probably dozens of varieties to choose from in your area, so ask for them at your local nursery.

In addition to water savings, there are other advantages to creating a xeriscape garden. For example, xeriscapes beautify, add fragrance, and enhance your home's value. Maintenance costs are reduced, and less power mowing means less gasoline is used, thus reducing exhaust and noise emissions.

One case in point: a Denver resident completely re-landscaped a yard that included 1/4 acre of bluegrass lawn. The use of xeriscape principles resulted in a 60-percent reduction of turf grass, cut water use in half, and saved many hours of mowing, weeding and fertilizing.

While a xeriscape yard is truly the ideal solution for a water-conscious family, do not be misled: quality conversions cost from hundreds up to tens of thousands of dollars. Yet some communities, in an effort to promote water conservation, offer rebates for replacing lawns with water-conserving plants. Check with your local water department.

Water Smart Savings: 30 to 50 percent.

Wild Grasses

One often overlooked rule of smart lawn management is to plant only grasses that can naturally flourish in your climate. For example, if your Rocky Mountain home town receives only 13 to 15 inches of precipitation annually, why plant a bluegrass lawn that needs an additional 24 inches of water per year? Instead, investigate other types of grasses like tall fescue, crested wheat and buffalo, which require up to 60 percent less water, and perhaps little or no sprinkling. As an unexpected bonus, native grasses also generally require less mowing than nonindigenous varieties.

One overlooked rule of smart lawn management is to plant only grasses that can naturally flourish in your climate.

Water Smart Savings: Up to 100 percent.

To Plant A Tree

Trees and shrubs, common to both xeriscapes and traditional landscapes, offer many advantages. Their broad leaves provide shade and lower the nearby air temperature, act as sound barriers, provide food and shelter for birds and other wildlife, absorb carbon dioxide and supply oxygen.

While a row of evergreens blocks winds that evaporate water, for instance, the air temperature under a shade tree may be 12 to 20 degrees cooler than the ambient temperature.

If that's not enough, the roots of bushes, shrubs and trees grow wide and deep, which stabilizes the soil and prevents erosion. Best of all, because the roots reach deeper into the soil, they generally require much less water than grass – sometimes up to six times less.

Water Smart Savings: 2,000 gallons per year for each 10 x 10-foot section of grass replaced with trees or shrubs.

Switch Hitting On The Farm

During water-stressed times, farmers switch to crops that require less water. Not long ago, alfalfa covered 75 percent of the cultivated land in Cuyama Valley, California; it now takes up only 25 percent. Replacing the alfalfa are carrots or nut trees, which require only half the water.

On a 100-degree day, a three-inch layer of mulch can reduce soil temperature by up to 25 degrees.

If you like to grow your own vegetables at home, contact your area U.S. Department of Agriculture office for information on water-efficient crops that will do well on limited water in your area. And while you're at it, apply a layer of mulch at least two to three inches deep over the root area of the plants helps to reduce soil temperature and retard moisture loss. On a 100-degree day, a three-inch layer of mulch can reduce soil temperature by up to 25 degrees.

Water Smart Garden Tactics

To sum up this section on lawns and xeriscape alternatives, following is a list of the most important water-saving practices you can adopt. With smart planning and a timely implementation, your yard's makeover from thirsty to thrifty should be an enjoyable and rewarding process. Best of all, it's long-term appeal will be tremendous.

1. Plan and design. Contact information sources such as local xeriscape groups, your city water department and nearby garden center for suggestions.

2. Replace thirsty lawns with trees, shrubs and native grasses.

3. Create watering zones by isolating high- and low-consumption plants into groups.

4. Remove unwanted weeds, which steal water from desired plants.

5. Test your sprinkler system – and your soil – and water only as needed.

6. Install a timer to control your sprinkler system rather than relying on memory.

7. Gradually phase out water-hungry plants.

8. Mulch the root areas around plants and create dirt watering basins or berms around tree trunks.

9. Don't water if the weather forecast calls for rain.

Drip Systems

Drip irrigation is the commonly used term for precise, low-volume watering. Trees, shrubs, gardens, and most ground covers – everything but the lawn – are economically watered with drip-type irrigation, because water placement is more precise than with sprinkling or spraying.

Drip watering can deliver 98 percent of the water where it's needed most – at a plant's root system. And a properly designed and operated irrigation system can reduce water consumption by 70 percent or more.

In addition, according to the American Water Works Association, 85 percent of all landscape problems are directly related to overwatering. And wouldn't you know it, most over-watering occurs when water is spread into the air.

Water The Earth, Not The Sky

Two natural processes return moisture to the atmosphere: evaporation and transpiration. In the irrigation industry, these principles are sometimes referred to as "ET" or evapotranspiration. Whatever term is applied to the phenomena, spraying or sprinkling water into the air is really the process of buying water, then wasting it.

A drip system is comprised of any method that delivers water precisely where you want it, as opposed to a sprinkler system which disperses water in a more general, scattered pattern. Drip setups allow us to accurately control how much water we use and where we put it while minimizing runoff, overspray and ET.

A drip system directs water precisely where you want it – and your plants need it.

Anatomy Of A Drip

The drip system consists of three main parts:

1. The adapter head, which attaches to a standard sprinkler head or water pipe and includes a flow regulator and filter.

2. A transmission system of pipes or hoses.

3. Individual emitter heads to control the direction and volume of water flow.

Planning and installing a drip system can take some time, but it isn't particularly labor intensive. Anyone who has enough technical expertise to use a shovel and pair of pliers should be able to handle it.

A drip feed system can be installed on standard sprinkler posts, such as this Saturn 4 unit from DISCO.

The Drip Irrigation Supply Company markets a system of drip components known as Saturn 4, which use sprinkler head or garden hose adapters to distribute water to an assortment of tubes and soakers. Available in various flow rates (from 0.5 to 20 gallons per hour), the different components and kits available cost from $6 to $20 and serve as a good overview of the whole drip-irrigation game. Consult a gardening store or DISCO, P.O. Box 42040, Las Vegas, NV 89116; 800-433-0646.

Water Smart Savings: Up to 70 percent.

Stay Out Of Jail Free

Some water-stressed communities have enacted temporary laws to help curb outdoor water use during drought times. One rather common regulation limits residents to watering by hand with a sprinkler can or pail, or by means of drip systems. Needless to say, proclamations of this sort are what put drip systems on the map – fast.

Other areas rely on even-odd watering days to keep residents in check. Rather than getting married to a calendar, the best solution here it to invest in an irrigation timer such as those discussed later in this section.

Water Smart Savings: A trip to court.

Planning A Drip System

A good way to begin planning for a drip system is by deciding which plants you need to water, and what volume or frequency of water they will require. This will determine what type of emitters you need, which in turn will determine the rest of the plumbing. In other words, figure out what results you want and build a system to suit.

Start by making a map of your yard on a large sheet of paper, indicating where each type of plant is located. Sketch your home, driveway, walkways, and all permanent features. Locate big trees, little ones, low and high shrubs. Also note the location of all water sources, including faucets, sprinkler heads, gutter downspouts, and even areas where storage cisterns may be kept.

Basic Drip System Components

Following is an overview of the components in a big-league drip system. Chances are good that you'll run across all of these at one time or another once you commit to drip irrigation. The apparent complexity of drip systems need not spoil your appetite for a water-efficient garden, however; once a system is purchased and correctly installed, it should be as reliable as an anvil.

The apparent complexity of drip systems should not spoil your appetite for a water-efficient garden.

CLOCK OR CONTROLLER Regulates the frequency, starting time, and duration of an irrigation system. It can be a simple clock timer or an expensive computer-controlled programmer, or something in between.

WATER VALVES Electric shutoffs that turn on and off the water to irrigate distant points; operate at a signal from the controller.

WATER REGULATORS Restrict the amount of water pressure delivered to the drip system. Since most homes have two to three times the water pressure needed by drip systems, these are necessary to keep from damaging the system components.

FILTERS Required to screen out sediments so that tiny emitter orifices operate without clogging. A must for grey-water systems.

DELIVERY PIPES Required to get the water from the source to the emitters. Most commonly made out of plastic, delivery pipes are available in stiff PVC (polyvinylchloride or "vinyl") or flexible PE (polyethylene or "poly") materials.

MICRO TUBES Flexible mini-distribution tubes that connect the main delivery pipes to individual emitters serving individual plants. Cost is a few cents per foot.

EMITTERS

– *Drip* These operate at a many different pressures, depending on need. Drip emitters are the most precise and efficient way to deep-root water. Different types are available to serve different functions.

A single sprinkler head can provide water for planting beds, trees and ground cover.

– *Misters and mini-spray sprinklers* These operate with low pressure and deliver low flow rates up to 0.75 GPM. They can aim water above ground over a fairly wide area, depending on type. Very thrifty and good for ground cover.

– *Porous and perforated pipes* These "ooze tubes" may be used above or below ground. They seep water slowly or quickly depending on need, and operate on 10 PSI or less. These pipes are just right for grey water distribution.

BUBBLERS Devices that can dispense a fair volume of water directly over the root zone, but at a reduced force to guard against erosion. Some types can be screwed onto existing sprinkler spray heads, some attach to hoses.

WATER WAND An extended applicator that can deliver water gently at the base of plants, while minimizing erosion and wasted water. From $15 to $20.

FERTILIZER INJECTORS An efficient way to feed plant roots while drip irrigating. These are especially useful for trees or other plants with deep root structures.

Inexpensive Kits That Save

Several name manufacturers offer kits that include emitters, tubing and adapters so you can run low-volume outlets to planting beds, trees and ground cover from a common sprinkler head. A simple kit might sell for $25, with fertilizer and low-volume sprinklers as options for $10 to $15.

One maker of such kits is Rain Bird, the company that brought you all those afternoons of running through impulse sprinklers as a kid. Today Rain Bird's DIY system promises a new kind of water-smartness with pop-up sprinklers, spray heads, timers, moisture sensors, and drip components. For more information, see your garden supply store or contact Rain Bird at 145 No. Grand Ave., Glendora, CA 91740; 800-247-3782.

Rain Bird provides a dozen product categories including timers, sensors, spray heads, and complete drip systems, all designed to conserve water.

Going Automatic

The purpose of this book is to help make you water smart, not a slave to water. So, for those tackling a Ph.D. in watering the yard and garden, nothing will make you feel smarter than a little self-contained irrigation microprocessor. Powered by either household current or batteries, this device can fulfill your every watering plan with only the slightest bit of attention on your part.

The purpose of these controllers, of course, is to relieve the forgetful human species of the repetitive and tedious task of watering plants on time and in sequence, day by week by month. And as we all know from experience, tedium does not well serve the spirit.

RainMatic sells a couple of different battery-powered irrigation controllers that work fine for a relatively simple garden, say with no more than two major systems involved. Consisting of a weatherproof housing, a microprocessor and a key pad, the RM2500 ($55) and RM3500 ($70) controllers attach between your garden faucet and irrigation system. Once programmed, the system turns the water on and off as desired over a seven- to 14-day cycle, depending on model. The RM3500 also has a "mist cycle" that can run for as little as three seconds to care for delicate flowers or seedlings.

Further information is available from RainMatic Corp., 828 Crown Point Ave., Omaha, NE 68110; 800-228-3615.

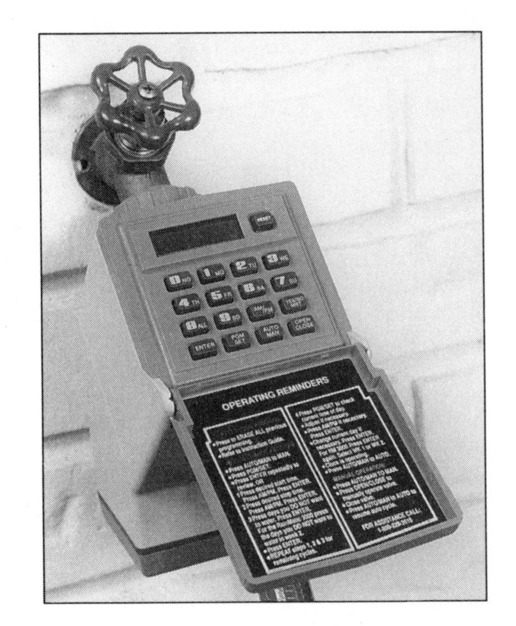

RainMatic sells a couple of different battery-powered irrigation microprocessors that work fine for a relatively simple garden. Costs are $55 to $70.

When The Rain Comes

While we're on the subject of automation, you should know that devices are also available to automatically shut down your irrigation system if rain falls or if the soil is too moist to need watering – another water savings. Actually, every good automatic irrigation system uses these moisture sensing devices today to maximize the system's accuracy and efficiency.

These sensors are buried in the soil, and if the soil is wet they prevent the water valve serving that area from turning on. Several styles are available depending on application. They can be used for almost any purpose: hoses with portable sprinkler heads, drip or mini-spray networks, or permanent sprinkler systems. RainMatic's own version, the RainSensor, plugs into the RainMatic controllers. The cost is $33 per sensor.

Be A Leak Detective

The more complex an irrigation system becomes, the more likely it is to leak. When this happens, it can happen big: one pinhole leak in a water pipe or faucet can spill out 170 gallons in 24 hours, about the amount of water a typical family uses in a day.

So while you're snooping around, remember to check hose connections for leaks too. Without your knowledge, hundreds of gallons can leak away while you're watering. Fortunately, aboveground irrigation leaks are just about the simplest things in the world to fix. Rennovating hose connections or even a faucet is as easy as turning off the water, disassembling, and replacing defective washers.

Water Smart Savings: Up to 5,000 gallons per month.

Water Storage

Every gallon of rain water that can be saved is one less gallon of water you won't have to buy from the city or pump from a well. So why let rain water go down the gutter?

The simplest way to store rain water is by using an ordinary trash cans as a cistern. Placed under a modified downspout, a can will salvage water from your rooftop to feed plants outdoors later on. Water storage can be as easy as this or as complex as you like, yet the principal stays the same: saving from a rainy day.

Catching Water Outdoors

The roof of your home can be made into a wonderfully lucrative watershed. How lucrative? *Sunset* reported that a 2,000-square-foot roof can shed up to 1,500 gallons from an inch of rain.

To store captured rain water you can use low-technology cisterns – a storage concept that has been around since mankind began living in caves. Then, while the rain is pitter-pattering overhead, you can be smug in the knowledge that your gutters are funneling this precious precipitation into cisterns and rain barrels for later use.

This 35-gallon rain barrel from Gardener's Eden captures water from a shortened downspout. The cost is $65.

An ordinary outdoor trash barrel positioned at the downspout – with an appropriate hole cut in the lid to accept the water – makes a perfect cistern. When you please, you can siphon the water out with your garden hose or install a hose spigot in the side of the container. To transport water uphill, you can use a modern submersible pump or dip in a watering pail for watering delicate flowers.

The best way to maximize your catch is to position one or more barrels at each downspout. *Sunset,* an excellent source of gardening information, also reported that one homeowner captured over 220 gallons in a series of containers after less than one inch of rain.

Once a rain collection system is set up, it should serve your needs reliably for years to come.

Setting up such an involved system will likely mean modifying the downspouts, finding a permanent home for the rain barrels, and figuring out how to use all that water you're going to collect. The good part is that once a system is set up, it should serve your needs reliably for years to come.

For the non-do-it-yourselfer, the Gardener's Eden catalog sells one such dedicated barrel for $65. The 35-gallon receptacle has a hole in the top for the downspout, along with a handy drain tap positioned part way up the side. A 50-gallon unit is also available. Contact Gardener's Eden at P.O. Box 7307, San Francisco, CA 94120; 415-421-4242.

Water Smart Savings: Up to 35 gallons per downspout.

Storage Precautions

Although rain water is essentially clean and appropriate for outdoor purposes, it can present some hazards, especially when stored long-term in outdoor containers.

Roofs may be contaminated by bird or wildlife deposits, decayed leaves, airborne pollutants, and in some cases materials used in the construction of roofs, such as oil and tar, and perhaps even asbestos. But even in the worst case, cistern water collected from roof tops should be acceptable for the irrigation of non-food source plants. However, one caveat: runoff from any new roof – asphalt or wood – may contain enough compounds to harm plants. Wait until a few good rainstorms have washed these chemicals away.

Other precautions should be observed no less seriously. In some parts of the country mosquitoes can breed in stored water. And in warm climates, heated cistern water can damage plants (cool city water or well water doesn't heat up as much as water stored above ground). So use light-colored barrels to reflect sunlight, locate the vessels in the shade if possible, and keep them tightly capped.

Capturing runoff from parking areas or sidewalks, even though feasible, presents a higher risk of contamination. This runoff is usually more polluted than roof runoff, and may contain higher levels of bacteria as well as hydrocarbons such as found in automotive motor oil.

Purifying Stored Water

The act of purifying stored water can be broken into four principal tasks: screening, settling, filtering and sterilizing.

Generally, screening and settling are sufficient to purify captured water for irrigation purposes.

The screening stage is simply the process of straining out large debris, such as leaves, twigs, and perhaps the occasional loose shingle. Filters may be made out of screen or coarse wire mesh, but they must also be cleaned periodically. Settling, the second stage, allows smaller debris to fall to the bottom of the cistern. Settling reduces the water's cloudiness and aids in the reduction of bacteria if the water is drawn a foot or so from the bottom of the barrel. Generally, these two stages are sufficient to clean the water for household irrigation.

Stage three, filtration, means percolating the water through mediums such as sand, mixed-media sand, ceramic or charcoal, or solar filters. The final filtering procedure involves disinfecting or sterilizing the water. Boiling sterilizes water, of course, and treating the water with bactericidal chemicals such as chlorine or iodine will disinfect it.

Stages three and four are generally considered necessary only to prepare water for consumption. Considering the universal availability of clean drinking water in North America, the need for these steps should be few and far between.

Commercial Cisterns

If you want to make sure you've got enough storage capacity to salvage every single drop of rain, heavy-duty tanks for both above- and below-ground installation are available from a variety of suppliers. They usually come complete with all necessary fittings, ready for installation. You'll no doubt need a big truck and backhoe, though.

A typical water tank is made from FDA-approved material for potable water and food handling, and often features seamless construction. It may also have an access hole on top for maintenance, built-in ultraviolet stabilizers, and a variety of inlet and outflow pipe options. Cistern sizes range from less than a hundred gallons to more than 1,500 gallons, although tanks that

Water storage containers come in all shapes and sizes. Big above- and below-ground cisterns suitable for home use are available from Norwesco, cost $312 to $520.

hold up to 10,000 gallons are also available.

Norwesco Products is one manufacturer of polyethylene cisterns ranging in size from 65 to 1,550 gallons. Above-ground model NW0550 holds 550 gallons and costs $358; NW0300 and NW0500 below-ground cisterns cost $312 and $520, respectively. For additional information, contact Hancor, Inc., P.O. Box 1047, Findlay, OH 45839; 800-537-9520.

On Grey Water Pond

Household water that's been recycled from bath tubs, bathroom sinks and washing machines is referred to as grey water, as opposed to black water – or raw sewage – which comes from toilets. Only grey water is generally considered suitable for use in the garden. It can be distributed over a plant's root system, which is the best way to irrigate shrubs and trees.

Collecting grey water can supplement your garden's water supply by about 200 gallons for each person in the household per week. For example, taking a shower or bath, which typically uses between 20 to 40 gallons, can easily add up to 100 gallons per week, and may even total twice that amount.

Dish rinse water and water from the laundry rinse cycle are other suitable sources of grey water, although some organic soap makers claim that grey water containing their products is fine for plants. All that's needed is a convenient collection and distribution system.

Some household water isn't suitable for irrigation. For example, laundry water containing harsh detergents, bleach and softeners, could damage plants. All is not lost, however: it may still be used for other cleaning purposes, such as washing patio furniture or scrubbing trash cans.

Soapy dish water likewise may be contaminated with too much grease and too many food particles, which can attract insects, clog soil and cause odors. Also, avoid water that has come into contact with diapers, infections, or that which has been used in the preparation of poultry.

Water Smart Savings: 100 to 200 gallons per week per person.

Handling Grey Water

Recently laws have been changed in some areas of the country to permit the use of grey water. However, the reuse of water does raise some health issues that should be addressed.

Primary concerns stem from grey water's diverse levels of quality and the potential for bacterial or viral contamination. The possibility of contracting any type of illness makes taking the right precautions the obvious step.

The safest way to distribute grey water is through an underground system, which helps to filter out harmful organisms.

Health officials do caution against handling grey water directly. The safest way to distribute it is through a drip irrigation system or leach (underground) application. Applying it below the surface helps to filter out harmful organisms.

Some grey water users prefer to catch the water in containers to provide better control over when and how it's used. But collecting and storing grey water in containers is also potentially risky because it increases the likelihood of developing bacteria and viruses. In some areas it's even forbidden by law, so check with local authorities.

When siphoning grey water, always use an inline hand pump or a submersible electric pump. And, need we mention, by all means resist sucking grey water through a hose to start a siphon action. By far the safest way to handle grey water is through a closed plumbing system. Again, check local building codes before setting up your own collection system. Making modifications to a home plumbing system usually requires both a permit and approval by a building inspector.

In summary: always use grey water the same day it's collected; apply only where it will not contact people; avoid letting grey water puddle or stand where it can't be absorbed quickly; do not spray or sprinkle it; and wear rubber gloves when handling grey water or its containers or plumbing.

Good Grey

If you plan to use grey water for irrigation, it's best to start with mild soaps and shampoos. Regular powdered soap may contain materials made up of sodium, whereas liquid detergents have few fillers and less sodium than powdered soaps. Detergents labeled as "biodegradable" are usually assumed to be the least harmful.

Laundry soap is available that has been specifically formulated for use in grey water systems. It contains a minimum of sodium, and has nitrogen, phosphorous and potassium, which are controversial, but beneficial to plant life. These soaps should be readily available in natural food stores.

Stay away from soaps containing bleaches, boron, borax and chlorine, and laundry softeners for grey water use.

Laundry soap is available that has been specifically formulated for grey-water systems.

Applying Grey Water

To use grey water effectively and safely, you may wish to observe these additional guidelines:

- Remember that water softeners often use sodium chloride. If you want to use water from your conditioner, make sure the conditioner uses potassium chloride instead.
- Before watering, make sure the grey water has cooled to at least ambient air temperature.
- Watch closely for any signs of plant damage. Some plants may show symptoms before others: salt-sensitive plants such as azaleas are most sensitive. Irrigate these with the purest grey water available.
- Any potted plant, indoors or out, needs fresh water only. Within the confines of a pot environment, alkalines and chemicals can build up to plant-wilting levels.
- If rain falls, shut off your grey water system so the rain can help to leach soapy buildup and chemicals from the soil.
- Avoid using grey water for fruits trees, vegetables, or plants that thrive in the shade.

Water Smart Savings: 100 to 200 gallons per week.

Spotting Soiled Soil

Even with careful grey water use, eventually small amounts of sodium or other harmful chemicals may begin to build up in the soil. This raises the soil's pH level. Rainfall helps to cleanse the soil, and you can occasionally help by judiciously sprinking with clear water. Flushing the soil a few times each year with a heavy soaking will also help to minimize residue build up.

To ensure that the sodium and pH levels of your soil isn't increasing to harmful levels, you can have your soil tested. Obtain lab references from your local gardening center or from a knowledgeable gardener.

The Collection System

A simple and effective grey water collection system can be hooked up to your washing machine, thus harnessing up to 40 or more gallons of water from each load of laundry. This setup utilizes a submersible pump to distribute water from a laundry room catch tank to whatever irrigation system you wish. Please refer to the laundry chapter for more complete information.

To collect water used in showers, tubs and sinks, you will need to tap into the existing household plumbing. Extra efforts are required to achieve a clean, workable system; getting a few contractor estimates wouldn't be a bad way to start.

Such a system needs several carefully engineered protective measures: a check valve between the drain lines and sewer to prevent "black water" sewage backup; and an overflow bypass into the sewer line in case the grey water system backs up. In addition, filters will be required to prevent clogging the drip system emitters.

If your grey water system will use a submersible or inline pump, be sure to get one that can lift water to distant irrigation points.

Once you have collected the grey water, you'll have to deliver it. So, if your grey water system will use a submersible or inline pump, be sure to get one that's rated for enough "head" – the height it can lift water – and adequate flow capacity to reach distant outlets.

Pools & Spas

A swimming pool holding some 20,000 gallons may seem like an excessive, even wasteful use of water. The fact is, however, that once filled, swimming pools and hot tubs are not always the water-squandering villains they're accused of being. In fact, an average pool uses less water annually than does a lawn of the same size. And a water-smart pool will consume only half that much.

Lawn vs. Pool

According to information released by the California Department of Water Resources, a 25 x 40-foot established lawn requires about 27,000 gallons of water during a typical year. As you've already learned, many homeowners exceed this amount by overwatering.

Well-maintained swimming pools are not the water-squandering villains they're often accused of being.

Now consider that a 15 x 30-foot pool, with a five-foot concrete border around it, occupies the same area as the 25 x 40-foot lawn. Losses for the pool could account for as much as 17,000 gallons per year, including backflushing the filter.

Water Smart Savings: Up to 10,000 gallons per year.

Figuring Pool Losses

Determining how much water your pool uses is easy and takes about a week. To begin, first mark the existing water level on the side of the pool with a grease pencil or dab of lipstick. Wait one week, then check the level again and measure the difference with a ruler.

To calculate your pool's weekly water loss, you'll need to first determine its surface area. You can do so by multiplying together the length and width (in feet) of a rectangular pool; by consulting with your pool blueprint or contractor to obtain the surface area; or by multiplying together the approximate length and width of an irregularly shaped pool. In one way or another, you should come up with an area in square feet. As an example, a 15 x 30-foot retangular pool has an area of 450 square feet.

Next, multiply the area of the pool by the amount the level dropped in one week (represented as a fraction of one foot), and you'll know how many cubic feet of water your pool has lost. Continuing with our example, suppose our 450-square-foot pool has lost an inch of water in a week. One inch, or one twelfth of a foot, is represented as 0.08 ft. Thus, 450 x 0.08 = 37.5 cubic feet of water lost in one week.

Finally, to convert cubic feet to gallons, multiple your figure for cubic feet by 7.5. In our example, 37.5 cubic feet x 7.5 = 281 gallons of water lost in one week. If this rate of evaporation continued for an entire year, the annual loss would be almost 15,000 gallons...plenty enough to save.

Cover Up

A few years back, a similar test was commissioned by a pool cover manufacturer to see how well a cover would arrest evaporative losses. Three nearly identical pools were built in close proximity to one another. One pool was left uncovered and unheated, the second was covered and heated to 87 degrees, and the third pool was heated to 87 degrees and left uncovered. After two weeks, it was discovered that the heated and covered pool had lost only 100 gallons compared with 680 gallons for the uncovered pool and 1,320 gallons for the heated and uncovered pool.

Thus, a solar pool cover provides many positive features and benefits extending beyond saving water. Heating costs are significantly reduced, as is the rather substantial cost of pool chemicals. Even the life span of the surfaces of a covered pool are greater than those of an uncovered pool.

Pool covers reduce evaporation water losses by up to 70 percent, which means you'll be running much less water through your water meter and into the pool. A cover also acts as an insulating barrier, sealing in precious heat. Working as a thermal collector, it can add five or 10 degrees to the water temperature. In some areas this is enough to eliminate the need for heating during summer months.

A cover protects the water from direct sunlight, which deteriorates the chlorine. This results in up to 50 percent less chlorine use. In the long run, a solar pool cover can also extend pool life by cutting down on the need for acid washes, liner

A pool cover can reduce evaporative water losses by as much as 70 percent, seal in heat, and keep your pool clean and safe when not in use.

repair, and filter, pump and heater replacement.

In addition to helping to reduce water, chlorine and heating costs, a pool cover makes for a safer pool. A final statistic: more than half of all the children who drown, do so in the family pool.

Water Smart Savings: Up to 100 gallons per day.

Choose Your Cover

Most covers must be hand-rolled, but may be rolled away on wheels to clear the perimeter area for use. Other covers, operated by electric motors, roll themselves in and out at the touch of a button. One of several automatic systems built by Poolsaver is hidden in a wooden housing that doubles as a seating area.

The ease with which a pool cover may be operated often determines how often the cover is used – and how much water is saved.

The ease of operation provided by such a system virtually guarantees a pool cover will get used, and of course water savings are the result. Fully automatic systems such as those offered by Poolsaver cost from $5,000 to $6,000. More information may be obtained from Poolsaver, 679 W. Terrace Way, San Dimas, CA 91773; 800-222-6837.

Pool & Spa Savings

Six primary water-saving strategies can serve pool and spa owners equally well.

1. Keep the pool covered when not in use.
2. Plug the overflow drain before using the pool.
3. Turn off the spray device on automatic cleaning equipment.
4. Clean filters by dismantling and cleaning rather than backwashing.
5. Backwash the filter system only when necessary.
6. Reduce the thermostat setting for heated pools. A lower water temperature will reduce both evaporative losses and energy use.

Water Smart Savings: 50 to 100 gallons per day.

Operation Splashdown

Here are three curmudgeonly pool suggestions we just cannot endorse: no water fights; no cannonballs; and no fun.

If you can't be a seahorse once in a while, what's the use of having a pool? Besides, most in-ground pools these days have cantilevered decking with jutting edges to help keep man-made tsunamis and tidal waves inside the pool instead of all over the concrete perimeter.

With all the water you're going to save with this section, you deserve a little bit of fun. Happy landings.

Auto & Garage

Automobiles and garages don't normally have water faucets, so how can they use water? Cars regularly need water for washing, and within the garage walls may also exist water conditioners responsible for a portion of our total home water use. Both subjects deserve inspection to achieve a truly water smart home.

EcoSource indicates that nearly two percent of our total household water applies to the care and feeding of automobiles. Additional water is used for the care and feeding of household water softeners, but how much depends on the type of conditioner and how much water your household consumes.

In this chapter we'll look at ways to save water when caring for your car, along with some economical alternatives to the traditional water-hungry water conditioner. Finally, some information on one newer development – the water circulation system – which can help to save an impressive ration of water in new or remodeled homes.

Automobiles don't normally have water faucets, so how can they use water? EcoSource says nearly two percent of household water applies to the care and feeding of our cars.

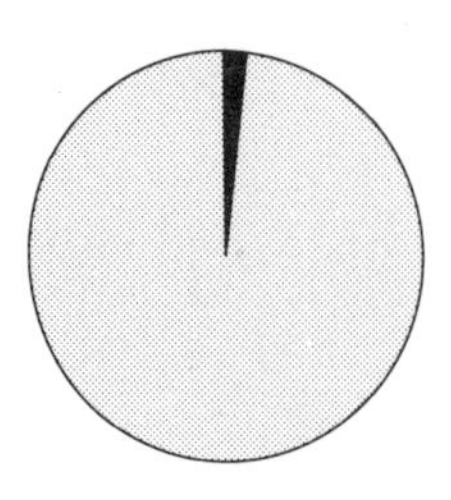

Auto & Garage
2%
of home water use

Car Washing

These days a decent car costs almost as much as a cheap farm in Kentucky, and what happens after we buy one? We let it sit outside in the dust, drive it on the freeway where it gets pelted with rocks and more dust, and then we park it outside at night to get covered with dew. Then we take our car to be washed (or do it ourselves), and the cycle begins again.

Is there an easier way to care for our cars while using less water? You bet. Following are some proven methods to reduce water use and keep your car cleaner, longer. And ways to use less water when it does need to be washed.

As an added benefit, your car should last longer, too. So keep that dream of a Kentucky farm alive.

Le Sponge And Bucket

There must be a dozen different ways to wash a car. Probably the most wasteful calls for turning on the hose and letting it run while we work our way around with a sponge or cloth. Since the average garden hose can flow between 10 and 16 gallons of water per minute, in the worst-case scenario a ten-minute car wash will use some 100 to 160 gallons of water.

The water smart way to wash: 10-quart bucket, adjustable spray nozzle and sponge gets the job done with about five gallons.

Mind you, that's the worst-case scenario. You're not really likely to turn the hose on full blast. Or are you?

On the more efficient side, a careful job with bucket and adjustable spray nozzle uses only about five gallons of water. The procedure: First wash off what dirt you can with a light spray from the adjustable nozzle. Then fill a 10-quart pail with water and whatever kind of soap you want to turn loose in the environment – Simple Green concentrate works well. Now get washing, starting on top and working down. Do the tires and wheels last. Rinse again with a light spray and dry in your own fashion.

If you had spent ten minutes on the job with the average garden hose turned on only halfway, you would have still average a kingly 66 gallons of water. This way you've used not even a tenth that much.

Water Smart Savings: 95 to 155 gallons per wash.

Avoid Dew Like The Plague

If you want dirt to stick to your car, there isn't a much better recipe than a day's worth of dust marinated in an overnight dew. That's why one of the best things you can do to keep your car clean and to use less water is to keep it under cover at night. This means under a car cover, beneath a carport, or in a garage.

Our experience tells us that a car kept out of the dew at night needs washing about half to one fourth as often as one that lives outdoors around the clock. Of course, actual water savings will depend on the method of washing selected.

Water Smart Savings: Two to four car washes per month.

Waxed Cars Are Easy

A waxed car stays clean longer than an unwaxed car. That's because a good coat of wax acts as a slippery shell over your car's body, making it harder for dust and dirt to adhere. The not-so-obvious payoff is that a waxed car will stay clean substantially longer than an unwaxed car, requiring less washing and less water.

How many car washes and how much water you will save depends on the wax job and the conditions to which your car is exposed. Our own tests show that an unwaxed car can pick up a

The smooth surface of a waxed car repels dirt just as it repels water in this photo, so the car stays cleaner, for longer.

Many automakers now offer clear-coat finishes that eliminate the need for waxing, thus reducing the need for washing.

good coat of dirt in a week's time, whereas a well-waxed car stays clean a hundred percent longer – two weeks, all told – in the same environment. Plus, at bath time dirt is easier to clean off a waxed surface. How often to wax a car? Turtle Wax generally advises four times per year.

New-car buyers take note: many automakers now offer urethane or fluorine clear-coat finishes that eliminate the need for waxing for several years. Just as with a waxed car, less dirt adheres to the new car's clear-coat surface, which of course reduces the need for washing.

Water Smart Savings: Two car washes per month.

Recycle At The Car Wash

Some commercial car washes are beginning to advertise that they use recycled water...as well they should. Whereas a typical car wash uses between 15 and 45 gallons of water per car, our sources tell us that with a recycling system at work, only 5.5 gallons of water is lost per wash – a reduction of 10 to 40 gallons.

Accountants also see the benefits: one car wash that we researched saved an average of $315 per month in water costs after installing a water-recycling system.

Water Smart Savings: 10 to 40 gallons per wash.

Water Systems

In the garage are found water heaters, water conditioners, and occasionally water circulation systems in newer homes. While water heaters perform a priceless service for us, they are in effect just servants of the rest of our home water system. There are no noteworthy ways to save water in the selection or use of a typical hot water heater.

Instead, let's delve into water conditioner systems first, since their use is widespread and recent improvements can result in substantial water savings. The technically interesting water circulation systems follow at the end of this section.

Backflushing causes traditional water conditioners to be big water users, with 90 or more gallons needed for every regeneration.

Anatomy Of A Conditioner

Millions of people enjoy the benefits of home water softening or conditioning systems, and it's not hard to figure out why. When softened water flows from the faucet it's free of so-called "hard" minerals that can be felt on the skin, that discolor sinks and tubs, and that cause scaly or granular deposits in our water heaters, cookware and appliances.

Inside the conditioner, acid and alkaline minerals are removed as water passes through a plastic resin screen system. Eventually, of course, all available resin surfaces are used up, so salt must be periodically introduced to regenerate the system. After the salt has loosened the deposits from the screens, the system is cleansed by backflushing with water.

This backflushing process is what causes traditional conditioners to be big water users, with 90 or more gallons of water being needed per regeneration.

Thrifty Conditioners

High technology has finally changed the water conditioner just as it has affected virtually every other aspect of our lives. Instead of using a Colorado River's worth of water for every regeneration, efficient new systems such as the Rayne RHEII regenerate themselves by using as little as 15 to 30 gallons.

In addition, the operation of these new systems is controlled by a water flow meter rather than by a timer. This means that the systems regenerate on the basis of water used instead of on a preset daily or weekly schedule. The simple advantage: the water conditioner regenerates itself only when necessary...and not while you're on vacation.

Various manufacturers provide these water-smart systems on a national basis. For information on the $1,049 Rayne RHEII,

Thrifty water conditioners such as the Rayne RHEII save plenty of water during regeneration: 60 or more gallons per cycle.

contact Rayne Water Conditioning, 800 Miramonte Dr., Santa Barbara, CA 93109; 805-966-1695.

Water Smart Savings: 60 or more gallons per cycle.

Potassium Is For Plants

There is a relatively simple way to harness the water used and discarded by water conditioners, and that's to use it in the garden. However, the sodium choride commonly used in water conditioners is distinctly bad for the soil and plants.

Here's a solution. Change from using sodium chloride to potassium chloride in your water conditioner, then tap into the waste water outlet on your conditioner and feed the water to a sprinkler or drip system, or to a backyard cistern for later use. The plumbing is not difficult, but you should check local regulations before proceeding.

Be aware also that potassium chloride is generally regarded as the second choice for providing soft water for the home. Consult your local dealer for recommendations pertaining to using potassium chloride in your own conditioner.

Water Smart Savings: 15 to 90 gallons per cycle.

Some Like It Hot

If we knew how much water ran down the drain each time we opened a faucet and waited for hot water to arrive, we'd probably faint. Over a year's time, this volume of "waiting" water has to number in the thousands of gallons per household. Predictably, we are most likely to waste this water in the morning, when our hot water system has been out of circulation and the pipes are cold.

One solution may be found in installing a water circulation system. Such a system virtually eliminates "waiting" water waste by continually circulating hot water throughout the household. Consisting of a 110-volt water pump, a thermostat or timer, a check valve and a return line, the system continually cycles hot water from the water heater to the farthest faucet and back to the water heater. This ensures that hot water will be available as soon as this faucet – or any other along the way – is opened.

Obviously, a hot water circulation system requires extra energy to keep the water in the water heater plus that in the pipes warm, and to circulate it by electric pump. These costs can be reduced by using a thermostat switch to turn the system on only when the water temperature in the pipes drops to a certain level; or having a timer turn the system on just before you awake and off after you retire for the night.

How much water you'll save depends primarily on how long you currently have to wait for the hot stuff to flow. But if each member of a family of four spends five minutes a day waiting for hot water at 3.0 GPM faucets, 60 gallons per day will be saved with a circulation system: 21,900 gallons per year.

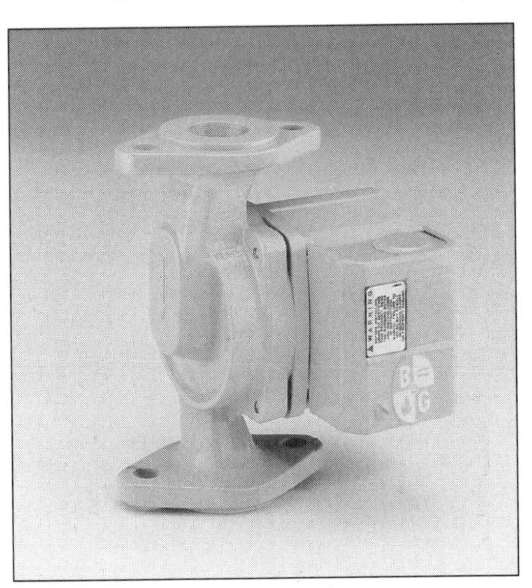

A circulation pump such as the Red Fox shown here continuously circulates hot water throughout the house, eliminating the need to wait for hot water.

The heart of the system, of course, is the pump itself. ITT Bell & Gossett's Red Fox SLC-30 is just one of the various circulation pumps available for home use. Designed for simplicity and reliability, the Red Fox sells for under $150; but don't forget to factor in costs for installation and a thermostat or timer. To learn more, contact ITT Bell & Gossett, 8200 N. Austin Ave., Morton Grove, IL 60053; 708-966-3700.

Water smart savings: 60 gallons per day.

Acknowledgments

This book was made possible by the widespread support and encouragement of local businesses, government agencies, friends and family members, product manufacturers, and most especially by the countless individuals who shared their valuable information soley by telephone.

The authors and publisher wish to thank the following individuals and organizations:

John Codero, Debra Dalton, Jacques DeBra, Debbie Ellwood, Lily Guild, the Office of California Senator Gary K. Hart, Russell Hodin, David Ingalls, Bruce Martin, Ken Rohl, Phil Schilling, and Robert and Suzanne Stein.

American Water Works Association, Association of Home Appliance Manufacturers, Atlanta Chamber of Commerce, California Department of Water Resources, City of Austin, City of Santa Barbara, Denver Water Department, Desert Water Agency, Goleta Water District, The International Society of Arboriculture, Los Angeles Municipal Water District, Metropolitan Water District of Southern California, Montecito Water District, Santa Barbara Public Library, Santa Barbara Water District, Santa Clara Valley, Texas Water Commission, Tucson Water, and United States Department of Agriculture.

Ace Hardware Corporation, American Standard, Andi-Co Appliances, Inc., Asko, Inc., Biocompatible Products, California Appliances, California Spa & Pool Industry Energy, Codes & Legislative Council, Chronomite Laboratories, Inc., Columbus Washboard Co., Cover-Pools, Inc., Crane Plumbing, Drip Irrigation Supply Company, Inc., Educated Car Wash, Everpure, Inc., Gardener's Eden, General Electric Company, Hancor, Inc., Holiday Inn, Home Improvement Center, Hope 'n Hagen's, Incinolet Products, Ingalls Plumbing, In-Sink-Erator, ITT Fluid

Technology Corporation, Kohler Co., Maytag Company, Miele Appliances, Inc., Mor-Flo/American, National Kitchen & Bath Association, NuTone, Poolsaver, Procter & Gamble Company, Rain Bird National Sales Corporation, RainMatic Corporation, Rayne Water Conditioning, Real Goods, Record-Wheatley, Inc., Resources Conservation, Inc., Richard S. Dawson Co., Santa Barbara Plumbing Supplies, Sears, Roebuck and Co., Simer Pump Company, Soilmoisture Equipment Corp., Sparkletts Drinking Water Corp., Specialty Photographic Laboratories, Turtle Wax, Inc., Unique Landscape Necessities, Universal-Rundle, Water Facets, Western States Manufacturing, Corp., Water Control International, Inc., Water, Inc., Whedon Products, Whirlpool Corporation, and White Consolidated Industries, Inc.

Bibliography

SELECTED REFERENCES

BATHROOM

American Water Works Association. "Faucets, Flows & Flushers" (September 1989).

American Water Works Association. "Yes, You Can."

California Department of Water Resources, Office of Water Conservation. *The Residential Retrofit Program—Retrofitting Household Fixtures to Save Water.*

California Department of Water Resources. *The Retrofit Way to Save Household Water* (January 1988).

California Department of Water Resources. "Water Conserving Plumbing Fixtures" (May 1988).

California Energy Commission. *Directory of Certified Showerheads and Faucets* (October 23, 1989).

Stu Campbell. *The Home Water Supply.* Garden Way Publishing, p. 199.

City of Santa Barbara, Water Conservation Program. "Toilets: Fix Leaks & Reduce Flush."

Consumer Reports. "How to Save Water," July 1990, pp. 465-473.

Department of Housing and Community Development, Division of Codes and Standards. *Low Flush Water Closets, Urinals and Flushometer Valves.* Information Bulletin SHL 90-01 (February 14, 1990).

1990 Earth Day fact sheet. "Is a Burger Worth It?"

EcoSource. "Water Facts" (Spring/Summer 1991).

Family Living. "Creating a Water Conserving Bathroom."

Kitchen & Bath Specialist. "The Nuts and Bolts of Bathroom Fixtures and Accessories" (November 1990).

Kitchen & Bath Specialist. "Water Conservation Device Wins New Product Award" (November 1990).

Remarks by R. Bruce Martin of Water Control International, Inc. Hilton Hotel, New Orleans, LA (October 30, 1988).

Montecito Water District. "Water Conservation Ideas."

New Age Journal. "Quick Fixes," pp. 56-57 (March/April 1990).

Office of Water Conservation, Denver Water Department. "Retrofit: Do it Yourself...and Save."

Omni Products. "U.S. Water Use by Faucets, Showerheads and Toilets."

Palmetto Paddlers Newsletter.

Plumbline. "Toilet Terminology."

Record-Wheatley, Inc. "All You Need to Know About Low-Flow."

Santa Barbara News-Press. "Unique Commode of Travel" (August 4, 1990).

Sylvia Porter's New Money Book for the 80's. Avon Books, New York: 1980, pp. 284-285.

Supply House Times. "How Flushmate Works," November 1988.

Sunset. "Bringing Your Toilet Up to Date."

Tucson Water. "20 Ways You Can Save Water."

Urban Conservation Measures. *Water Conservation Reference Manual,* pp. 15-16 (March 1984).

KITCHEN

Association of Home Appliance Manufacturers. *Water Consumption Survey of 1987 Clothes Washer Models.*

Association of Home Appliance Manufacturers. *Water Consumption Survey of 1989 Dishwasher Models.*

Consumers Digest. "Water Softeners," March/April, 1990, pp. 42-43.

EcoSource. "Water Facts" (Spring/Summer 1991).

Technical Bulletin, Maytag Consumer Education Department. "Energy Conservation—Dishwashers," Form No. 316YG-0191.

LAUNDRY

California Department of Water Resources. "A Checklist of Water Conservation Ideas for Laundries and Linen Suppliers."

Stu Campbell. *The Home Water Supply,* Garden Way Publishing, p. 199.

Consumer Reports 1990 Buying Guide Issue.

EcoSource. *Water Facts* (Spring/Summer 1991).

Maytag Consumer Education Department. *Automatic Washer Buying Guide.*

Maytag Consumer Education Department *Technical Bulletin.* "Energy Conservation—Clothes Washers."

Maytag Consumer Education Department, *Technical Bulletin.* "Suds-Saver Washers," Form No. 301YG-1290.

Maytag Consumer Education Department *Technical Bulletin.* "Water Conditions and Conservation," Form No. 340YG-0190.

Sylvia Porter's New Money Book for the 80's, Avon Books, 1980, pp. 286-287.

YARD, GARDEN & POOL

California Department of Water Resources. *Landscape Water Conservation Guidebook No. 8* (March 1988).

"California's Environmental Struggles," *The Washington Spectator,* August 15, 1990, p. 1.

Stu Campbell. *The Home Water Supply,* Garden Way Publishing, pp. 204-208.

City of Santa Barbara, Water Conservation Office. "Lawn Watering Guide."

Denver Water Department, Office of Water Conservation. *Xeriscape Colorado: Source Directory for Un-Thirsty Plants* (1987).

"Drip." *Sunset,* July 1988, pp. 68-76.

"Drip: Watering the Home Garden." University of California Cooperative Extension, No. 7107.

"Drought Survival Guide for Home and Garden." *Sunset,* May 1991, pp. 1-31.

EcoSource. *Water Facts* (Spring/Summer 1991).

"Good looking...unthirsty." *Sunset,* October 1976, pp. 78-87.

Robert Kourik, *Gray Water Use in the Landscape* (1988).

Art Ludwig, *Greywater Information* (1991).

Miracle-Gro. "20 Ways to Save Drought-Stricken Gardens."

Office of Water Conservation, Denver Water Department. *Plant Focus 1990.*

Raindrip, Inc. "Drip Watering Made Easy" (1983).

L. Ken Smith, *40 Ways to Save Water in your Yard & Garden,* Environmental Design Consultants—International (1977).

Santa Barbara City Department of Public Works. *Guidelines to the Approved Use of Greywater* (April 1990).

Sonoma County Water Agency. *Xeriscape* (1988).

State of California, Department of Water Resources. "Are You Using Gray Water During the Drought?" (April 27, 1977).

State of California, Department of Water Resources. *How to Produce a Lawn Watering Guide* (January 1987).

State of California, Department of Water Resources. *Captured Rainfall,* Bulletin 213 (May 1981).

"The Unthirsty 100," *Sunset,* October 1988, pp. 74-83.

"Using Household Waste Water on Plants," University of California, April 1977.

Water Conservation Products, Inc. *Water Conservation Handbook* (1990).

Water Education Foundation. "Landscape Design II."

Water Education Foundation. *Layperson's Guide to Water Reclamation* (1989).

"Water Wise." *Los Angeles Times,* August 19, 1990.

Xeriscape Colorado! *At Home with Xeriscape.*

"Xeriscaping Can Cut Demand for Water," *Western City*, December 1990.

AUTO & GARAGE

California Department of Water Resources. "A Checklist of Water Conservation Ideas for Car Washes."

California Department of Water Resources. "The Water Conservation Checklist."

EcoSource. *Water Facts*, Spring/Summer 1991.

Metropolitan Water District of Southern California. "Home Water Use."

Turtle Wax, Inc. "The Compleat Guide to Washing and Waxing Your Car."

GENERAL REFERENCES

Greg Cailliet, Paulette Setzer, Milton Love, *Everyman's Guide to Ecological Living*. New York: The Macmillan Company, 1971.

Wilson Clark, *Energy for Survival*. Garden City, NY: Anchor Books, 1975.

GAIA: An Atlas of Planet Management, New York: Anchor Press, 1984.

E.F. Schumacher, *Small is Beautiful*. New York: Harper Colophon Books, 1975.

BULLETINS

Denver Water Department. *55 Facts Figures & Follies of Water Conservation*.

Santa Barbara Bank & Trust. "26 Ways to Save Water."

Southern California Water-Energy Conservation Partnership. *25 Ways You Can Conserve Water*.

PERIODICALS

American City & County
Buzzworm
Consumer Reports
Earthwatch
Sierra
Sunset
Monitor
Time, August 20, 1990, pp. 58-61

JOURNALS

The Arizona Republic
The Arizona Sun-Journal
Los Angeles Times
The Santa Barbara Independent
Santa Barbara News-Press
USA Today

Index